WE,
HOMINIDS

WE, HOMINIDS

An Anthropological Detective Story

FRANK WESTERMAN

HEAD
of ZEUS

An Apollo Book

First published in Amsterdam in 2018 as *Wij, de mens* by Querido Fosfor
First published in English in Australia in 2021 by Black Inc. Books
First published in the UK in 2022 by Head of Zeus Ltd,
part of Bloomsbury Publishing Plc

9 7 5 3 1 2 4 6 8

A catalogue record for this book is available from
the British Library.

ISBN (HB): 9781803281520
ISBN (E): 9781803281506

Text design and typesetting by Typography Studio
Illustrations and maps © Yde Bouma

Printed and bound in Great Britain by
CPI Group (UK) Ltd, Croydon CR0 4YY

MIX
Paper from
responsible sources
FSC® C171272
FSC
www.fsc.org

Head of Zeus Ltd
5–8 Hardwick Street
London EC1R 4RG
WWW.HEADOFZEUS.COM

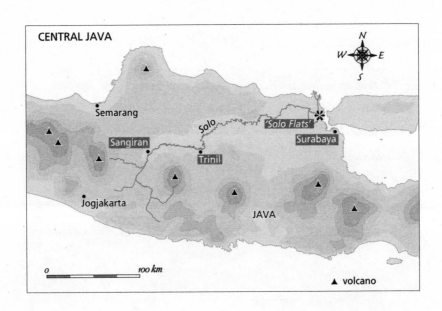

CENTRAL JAVA

Semarang

Sangiran

Solo

'Solo Flats'

Surabaya

Trinil

Jogjakarta

JAVA

0   *100 km*

▲ volcano

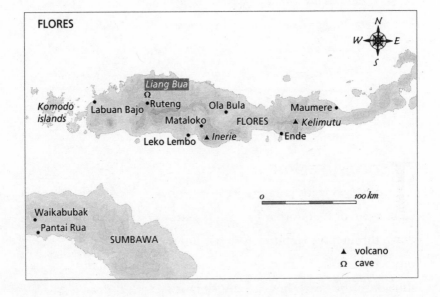

FLORES

Liang Bua
Ω

Komodo
islands

Labuan Bajo

Ruteng

Ola Bula

Maumere

Mataloko

FLORES

▲ Kelimutu

Leko Lembo

▲ Inerie

Ende

Waikabubak

Pantai Rua

SUMBAWA

0   *100 km*

▲ volcano

Ω cave

Prologue

*It took me a long time to find out where he came from. The little
prince, who asked me a great many questions, seemed never to listen
to mine. It was the words spoken inadvertently that, little by little,
told me everything... When he first beheld my airplane... he asked.*

'What kind of a thing is that?'

'It's not a thing. It flies. It's an airplane. It is my airplane.'

I was proud to have him know that I could fly. Then he cried out:

'What! You mean you fell from the sky!'

'Yes,' I replied modestly.

'Oh my! That's funny...'

*And the little prince started hooting with laughter. That irritated
me greatly. I prefer to have my misfortunes taken seriously.*

Then he added:

*'Well then, so you come from the sky too! So which planet are
you from?'*

ANTOINE DE SAINT-EXUPÉRY, The Little Prince

I T IS WHIT MONDAY IN 2012, and the single-engine Cessna
Skyhawk with tail number PH-SJK begins its descent over the
town of Ouddorp. For a full minute the pilot pushes the nose
down through the clouds, to catch a glimpse of the coastline and his
projected course. In the distance, banks of fog rise from the sea. They
slide onto the beach in one smooth motion – in the same way the
first marine animals must once have crept onto land.

Earlier that morning, at 10.22 a.m., the Dutch Meteorological Institute had issued a bulletin, warning pilots about the influx of moist air from the north-west. It is going to be a glorious day, but along the coast there is a 'chance of scud and frequent mist'.

The kind of mist the pilot sees rising up before him occurs, on average, once every two years and is known as 'sea fret'. The name stems from the optical illusion of boiling surf: the waves do not merely splash, they steam like kettles.

At 11.19 a.m. the plane reaches the bottom of its dive: 450 feet, the distance covered by someone with a large shoe size taking 450 steps heel-to-toe. Place that distance on end and you've got an imaginary ladder to the clouds, which you can then climb right up. It won't take long.

As soon as the pilot has eyed the landscape, he pulls back on the Skyhawk's joystick and climbs. Engine purring deeply in the background, he banks towards the open sea, hoping to fly around the upcoming mist. It is 11.20 a.m. (only one minute after the dip beneath the clouds) when PH-SJK establishes radio contact with the control tower at Rotterdam Airport. The pilot asks permission to land along the HOTEL approach, swinging in over the estuary of the rivers Meuse and Rhine, and then eastward above the New Waterway – the same route taken by the oil tankers and container ships, but higher.

The tower's code is PAPA. At its foot lies the runway, glistening like an unfurled carpet in the sun. It is all, quite literally, smooth sailing.

'This is Rotterdam information PAPA,' the air traffic controller responds. He tells the pilot to climb to 1500 feet and call in again above the town named Hook of Holland.

Considering its course, cruising speed and position on the radar, PH-SJK should be at the appointed spot within five minutes. But PH-SJK does not call again. Never, and to no one.

An inhabitant of Ouddorp is the last to get a 'visual fix' on the red-and-white Cessna: the man sees the tail of the four-seater dissolve into the clouds above the row of dunes.

AT THAT SAME MOMENT, A FEW MINUTES' flying time north of Ouddorp, a group of parents and children arrive for an outing on the *FutureLand Express*, a powerfully built train consisting of a farming tractor pulling two trailers converted into passenger compartments.

The passengers are elated; they are among the first who will be dropped on virgin territory. The newly reclaimed Maasvlakte 2 is still virtually untouched terrain. Physically speaking, the hop down from the running boards onto this new land-in-sea is only one small step; still, it feels like something important, like something that has to do with all of us.

From the concrete platform at the edge of the existing harbour grounds, the *FutureLand Express* is scheduled to depart at noon sharp. The morning sky is blue, but just before noon it clouds over. A breeze comes up, the temperature drops. The three-bladed GREEN-CHOICE turbines stir rather forlornly in the mist, mixing the silty sea air with the factory smoke from the harbours.

Earlier that week, Maasvlakte 2 went online with a fireworks display of blue smoke. The opening of the surrounding beach to the public – on Tuesday, 22 May 2012 – was celebrated with a champagne toast. That new beach is backed by a row of dunes fourteen metres high, and behind those are the 'sand works' for the quays to be built here soon. The planting of beach grass has been delayed, which means Maasvlakte 2 still looks like the Sahara.

But on this Whit Monday, a chilly mist shoos the swimmers and sunbathers home prematurely. The *FutureLand Express* is still on schedule, even though visibility shortly before departure has shrunk to fifty metres, one hundred at most, here and there. Neither the driver nor his passengers hear the drone of a single-engine plane. Only the screams of the gulls.

IT IS A RARITY, the disappearance of a plane in one of the most densely populated countries on earth. Flying itself, though, is not: aerial navigation has come to feel like second nature.

One year later, the Dutch Safety Board issues a 151-page report entitled *Plane Missing*. Unlike with the ships in the Bermuda Triangle, though, the disappearance of the Cessna Skyhawk is not permanent. It lasts precisely 301 minutes. Five hours after its final contact with Rotterdam PAPA, the wreck of PH-SJK is found 800 metres past the farthest stop-off of the *FutureLand Express* – close to the Princess Amalia and Princess Alexia harbours. Twisted and broken-winged, the plane lies beside its own impact crater, having made a quarter-turn. Like a bird that's hit a picture window. The speedometer needle is stuck at 118 knots, or 219 kilometres per hour.

The body of the fifty-year-old pilot is hanging in the front of the cockpit. The man who fell to earth is unconscious, but still breathing. He will not regain consciousness, and he dies two weeks later in hospital. The screen of his smartphone shows a series of missed calls. In his flight bag is a Visual Approach Chart for Rotterdam Airport. The map (printed in 2008) shows the coastline to be three and a half kilometres further inland.

The rescue team, too, works with maps that have recently become obsolete. The Cessna would have crashed in the sea had the coastline not been moved. The water has made way for dry land. But when PH-SJK's last known coordinates (thirty-six seconds before impact) were entered into the computer at Rotterdam Airport, the monitor showed a position at sea. Although the digital chart bears the note 'BEING RECLAIMED', the colour is aquamarine, indicating water.

'We have, we think, seen . . . [unintelligible] crash at sea,' an air traffic controller reports to the coastguard. Five rescue boats head out from the shore.

IF MAASVLAKTE 2 is a national sandcastle, the Princess harbours are its moats. The sand itself comes from an undersea quarry, sucked up by the dredgers' vacuum tubes from a sandbank six kilometres offshore. During the ice ages, that sandbank was on dry land. Wandering across the windy plain between what is now England and the Netherlands were hippopotamuses and hyenas, mammoths and rhinos, cave lions and forest elephants.

By excavating the floor of the North Sea, we hominids unintentionally do something absurd: we blow prehistory back to the surface. The dredgers, of course, are seeking sand and gravel, which they then 'rainbow' in great muddy arcs onto the spots where land is to be made. But their bycatch consists of mammoth molars, moose antlers, petrified hyena droppings – the remains of fauna from before the dawn of history.

In principle, two metres' visibility is enough for the passengers on the *FutureLand Express* – as long as they can see the sand beneath their feet. They are beachcombers. They are not looking for washed-up crates of whisky, nor for seashells either, but for fossils. For them, first prize would be a humanoid skull.

Further down the coast, in Zeeland, the first bit of primitive human has already been found. A hiker picked it up from amid the detritus spat onto the land by a cutter-suction dredge. It was a skull fragment with a ridge above the eye sockets that modern humans do not have. Further examination showed it to be the first Neanderthal ever found in the Netherlands; in 2009, under the name 'Krijn', it was put on display for the rest of the population. During its lifetime, it must have been a hunter wandering the delta of the rivers Thames, Rhine and Meuse, somewhere between 100,000 and 140,000 years ago.

IN THE MEANTIME, the icecaps had melted, the sea rose, the Neanderthals died out and the wingless *Homo sapiens* taught itself to fly. It also taught itself to separate dry land from the waters.

The commission for the new land has not yet been officially completed when the Cessna Skyhawk appears. The tower instructs the pilot to climb; instead, he dives under the clouds one more time to see where he is. Like a dolphin, but in reverse. There is one difference from just a few minutes earlier: beneath the clouds there now hangs a dense mist. The pilot continues to lean on the joystick. His plane submerses itself in a grey world of vapour that stretches all the way to the new land's surface, riddled as it is with fossils.

1

THE MORNING RUSH HOUR IS OVER; the first off-peak hour of the day has arrived. My intercity express is running five minutes late, which I see as a providential concession to my need for coffee and victuals.

I'm on the trail of a story, but this time I'm not alone. For the last few months I've been teaching at Leiden University, as a writer-in-residence. It was twenty-five years ago that I last sat in a university classroom; now I'm standing at the front of one. I am no longer a student; I *have* students. Forty-one of them, the list says – most of whom often don't show up. The rest, the hard core, will be joining me later, at noon. They major in Dutch, English, French, the humanities, education, art history, philosophy or theology, although the last of these no longer exists as such: it is now called 'religious sciences'.

My lectures are on the art of reportage. On my very first day before the class, I announced that we would be going out on the Meuse River. Today, after four weeks indoors, it's finally going to happen. Our meeting place and our point of departure is the station at Tegelen, on the river's shore in Limburg province. What we experience together this afternoon will be the subject of an article by each student, which will count towards their final grades.

The intercity express has arrived. In the aisle, at knee height, appears a golden retriever in a white leather harness. The animal is searching for a seat for an Indonesian-looking woman walking behind him, who holds the leash in one hand and a cane, which ticks back and forth, in the other. They settle in together: the golden retriever on the floor at my feet, his mistress with her shoulder bag on the orange-upholstered seat across from me.

There are three days to go before the Feast of St Nicholas, and I am eating gingerbread nuts. Two moist eyes regard me from below. The dog is wearing a thin, sleeveless smock, like an athlete's. Instead of a racer's number, though, it bears the text 'Don't Pet Me'. That means feeding must be taboo, so I put away the crackling cellophane bag.

'Thank you,' the woman says. By feel, but no less accurately for that, she pulls a sheaf of papers from her bag. Then, suddenly, her wristwatch starts talking: 'Am-ster-dam Cen-tral, nine-oh-seven.' It is our departure time, according to the timetable, but the blast of the conductor's whistle is long in coming.

I am off to work myself, so I pull out my notebook. It is a soft-cover black diary, and the flyleaf bears pre-printed text: '*In case of loss, please return to . . . As a reward: $ . . .*'. Not now. I turn the page and start scribbling something about the golden retriever, the gingerbread nuts, the talking wristwatch.

When I look up again, the woman across from me is reading. The pile of paper on her lap turns out to be an inkless book. Her fingertips scan the page in tandem, inwards and then back out again. Reading braille is a two-handed affair, a bit like knitting without the needles, or the wool.

I'm astounded by the connotations: reading blindly, flying blind, blind faith. By braille. By seeing-eye dogs. No, that's not exactly right. By the symbiosis between people and animals.

REPORTING, I TELL my students, is born out of a sense of wonder. Whatever you're reporting on, you could always precede it with a mental cry of 'Listen to this!' Sometimes your message is so urgent that you don't even have enough breath for that. 'We have won!' you shout then, if you happen to be a courier racing into Athens from Marathon.

I ask the students if any of them are runners. And, if so, whether they realise that their sport originated with the bringing of news. From courier to *corriere* to current to *krant*, the Dutch word for newspaper. Unlike the hare or the fox, we once started running because we had something to say.

What it boils down to is this: you have seen or heard or smelled something, and you want to pass this along to those who weren't there. Language is the medium. Twenty-six letters and a handful of punctuation marks, no more and no less than that.

Every reporter should feel like a child racing home to tell their parents about the exciting thing that just happened. That forces you to use short sentences, to establish priorities, first things first. But that is a method I've abandoned. Reality is too unruly to be in any hurry. Too crooked for cutting corners, too convoluted to iron out. It has too little rhyme or reason for a limerick.

'Tell your story to a child,' I say to my students. I go looking for a piece of chalk, but all I find is a marker pen. The blackboard behind me is, it turns out, a whiteboard. 'A child relies on more than just its eyes,' I continue. 'It knows that the most important things can't be seen.'

The classroom is musty. I ask someone to open a window while I hang up a map of Indonesia, the necklace of islands from Sumatra to Papua New Guinea. The erasers for cleaning the board, I find, have a magnetic backing – a stroke of luck.

Pacing back and forth in front of the Java Sea, I talk about my plans for the weekly tutorials between now, October 2016, and the Christmas holidays. 'We're going to begin at the beginning,' I hear myself say. By 'beginning', I quite literally mean 'from scratch'. I am on the brink of writing a long piece of reportage myself, but still have almost nothing to work with. Only the idea that I finally want to run to ground, in the wild, and the hunting grounds I will find it in.

I invite my students to help with the set-up. Working as a collective, we will chart out the major lines of the story, which means that together we will be present at the birth of a new book. This book.

In practice, I go on, we will adopt the methodology of a detective who has been assigned to a case: confirming hunches, interviewing witnesses, working out possible scenarios, thinking out loud about possible motives. And also forensic investigation at the scene of the crime.

Is there a body?

Yes, there is. 'LB1' is written on the tag on its toe.

'Have any of you heard of Flores Man? *Homo floresiensis*?' I am getting ahead of myself. Fortunately, though, I've already hung up the map. 'Or Flores, the island?'

According to the CIA's *World Factbook*, the Indonesian archipelago consists of 17,508 islands. Roughly a thousand of those are inhabited. As a nation, Indonesia is Islamic, but at least one island is predominantly Hindu (Bali), one is Protestant (Ambon) and one is Catholic (Flores).

My hand slides over the string of emeralds. 'Sumatra, Java, Bali, Lombok, Sumbawa, Flores,' I say. 'And here, between Bali and Lombok, you've got the Wallace Line.' Now I am starting to sound an awful lot like a geography teacher.

Neither the Wallace Line nor the man after whom it was named, Alfred Russel Wallace, ring any bells for my students. We are at the oldest university in the Netherlands, established in 1575, but my lecture series is for the Faculty of Humanities: not a single student from the 'hard sciences' has signed up. I back up and make another start, this time with Charles Darwin's *On the Origin of Species*, published in 1859.

I leave the Galápagos Islands and cut back to Indonesia, where Alfred Russel Wallace – independently of Darwin – simultaneously came up with the theory of evolution by natural selection. Wallace pointed to the deep marine trench between Bali and Lombok, on both sides of which divergent flora and fauna had developed. The islands west of the divide were characterised by a typically Asiatic biodiversity, and those to the east by a typically Australian one.

Flores lies to the east of the Wallace Line – that is all I am trying to say. 'And there, in 2003, in a cave, they found a skeleton.'

It was the remains of an adult female. Her skeleton was a primordial fossil. Although fully grown, the woman herself must have stood 1.04 metres tall. Her dwarfish stature would not necessarily have been a surprise, had it not been that her head was also small. Exceptionally small. Her skull was the size of a coconut. A grapefruit, some claimed.

The fumbling with pencil cases and the shuffling of papers stop. In the silence that descends over Classroom 0.04 of the Faculty of Humanities, I bring my students up to speed on LB1's unusual anatomy:

- Her cranial capacity was that of a chimpanzee (400 cubic centimetres). By comparison, we, *Homo sapiens*, have three times as much brain volume (1200 to 1400 cubic centimetres).
- Judging from the structure of her spinal cord and wrist joints, she was not a primate who lived in trees: LB1 walked upright, just like *Homo habilis*, *Homo erectus* and other hominids.
- Her feet were flat, well suited for covering great distances.
- Dating techniques indicated that she lived 18,000 years ago. Her descendants became extinct only 12,000 years ago, after a volcanic eruption covered Flores in a layer of burning ash.

'A new human,' the team of archaeologists had claimed. On 28 October 2004, this miniature human, who weighed barely 25 kilograms, appeared on the cover of *Nature*. Within a day, almost every news service in the world had 'welcomed' *Homo floresiensis* as the most recent scion in the family of hominids.

'Flores Man' may have constituted a new branch of the family tree of the species *Homo*, but no one knew which branch that was. What didn't make sense was that LB1 had no more cubic centimetres of brain volume than an ape, yet she was intelligent enough to make tools and to hunt (judging from the broken bones of animals and the stone axes scattered around her). That combination seemed preposterous. Flores Man was, in every respect, a *Fremdkörper,* an interloper.

To extend the metaphor even further: if all the hominids discovered to date had been varieties of fruit, LB1 was a Christmas tree ornament. A practical joke on the part of the Supreme Being.

How does this affect our view of what it means to be human?

Now I am getting close to my idea.

'DEVIATION VS NORM,' I write on the board. What is normal? And how do you determine that?

To me it is about more than just size, flat-footedness or brain capacity. Big/little, fat/thin – these are blatant, quantifiable contrasts. I want to gradually extend the line from the outward to the inward, from stature to behaviour, from then to now. What is abnormal? And who decides that?

These are questions that interest me. As a child, I was taught the virtue of propriety, of keeping one's head down. Rubbing up against the median was the way to play it safe – that was the home base it was best to keep within arm's reach. But now, barely one generation later, children are expected to do the very opposite. To stand out. Go for it. 'Differentiate yourself' has become an imperative, a calling. Differentiate from what? It makes no difference. That which makes you different from the rest, that is who you are. One step further, and the difference becomes the similarity.

We all slam our tails down hard on the water, but unlike the whale, we humans believe that our splashing matters. 'We tend to see ourselves as the measure of all things,' I tell my students. 'But why, in fact, do we call Flores Man "small"?'

In no time, my students succeed in reasoning away the concepts of 'large' and 'small'. If you consider something or someone small, that is relative to your own size. The elephant looks down on the mouse, the mouse looks down on the ant. That's right, I reply, but on the other hand, everyone starts off small. We're able to empathise with the dwarf because at one point all of us looked up to big people. We all clung to skirts and trouser legs. Everyone has held out their little arms; every child has been picked up hundreds of times by aunts and uncles – and just as often been put back down in their place.

I quote Antoine de Saint-Exupéry: 'All grown-ups were children first (but only very few remember that).'

The students are writing it all down.

'But what if LB1 just happened to be a small person?' The rebuttal comes from Lian, a junior in philosophy, herself Asian and rather small. ('That's why I wear heels,' she tells me later.) 'Or what if she was a midget? I mean, what if she was the exception?'

I like rebuttals. In the archaeologists' defence, I state that they had considered that possibility, investigated it and ruled it out. They had made their discovery in a cave by the name of Liang Bua. LB1 stood for Liang Bua 1. Beneath this rocky dome they also found the ribs and arm bones of other Flores Men: LB2, LB3 and so on, up to LB9. Although no other skulls were present, these bones belonged without exception to miniature people.

To make things even more fantastical: in the same cave lay the skeleton of an extinct species of stork. A colossus. Upright, this giant stork would have been almost twice the size (1.8 metres) of the Flores people. This too seems like a practical joke on the part of a divinity who is trying, with Satanic glee, to make people believe in fables: the bigger the stork, the smaller the babies it delivers.

Flores' fauna seemed specialised in anomalous formats. Halfway through the twentieth century, the Dutch missionary Father Theodor Verhoeven found on Flores the fossil remains of elephants with the shoulder height of a pony. These were not mastodons but stegodons: miniature elephants with mini-trunks and mini-tusks. Their calves would have been as easy to pick up as a toy animal.

The priest was the first to sink a shovel into the floor of Liang Bua. In 1950, in a corner of the cave where stalactites hung from the ceiling like runny candles, he had a test pit dug. There he found – the fairytale has just begun – skeletons of rats the size of dogs. Father Verhoeven lives on today in the generic name *Papagomys theodorverhoeveni*, otherwise known as Verhoeven's giant tree rat. He found long-nosed and stub-nosed varieties, ground-dwellers and tree-dwellers; today's water rats are minuscule by comparison.

Even today, Flores is host to various animals of gigantic form. In addition to the rats, the turtles and lizards on the island's beaches are enormous. The latter can be up to three metres in length, and move like long-legged crocodiles. These are the Komodo dragons: refugees from Jurassic Park, as ponderous as the tourists who come to see them.

On this island, animals assumed to be large in other parts of the world (elephants) are small. Conversely, the crawling things (turtles, lizards, rats) are super-sized. The topsy-turvy world of Flores: what better place to get to grips with divergence and norm?

'We are going down the rabbit hole,' I announce – and we will have no need of *Alice's Adventures in Wonderland* to do so. Flores is a Wonderland that actually exists.

What we will need, however, are guides. For starters, I plan to delve into the life story of Father Verhoeven. He perfectly matched the profile of the tragic hero. 'THE MAN WHO DIDN'T DIG DEEP ENOUGH,' I write on the whiteboard. Father Verhoeven seems perfectly suited to mediate between the idea we are hunting down (why do we see ourselves as the norm?) and the hunting grounds themselves (Flores as a treasure-trove of anomalous life forms). For the time being, he will be our pilot.

I still know next to nothing about Theodor Verhoeven, except that he must have been an extraordinary missionary: a cleric with a bent not only for the heavenly but also for the earthly and the subterranean. He carried out a series of excavations at Liang Bua cave in the 1950s and '60s, going deeper and deeper each time. What are the chances, after all, of turning up the skull of an unknown hominid species when you drive a spade into the ground in some far-off corner of the world? In some miraculous way, Father Verhoeven was spot on. The only thing was that his pit was only three metres deep, while the hidden treasure, the skeleton of LB1, was buried at 5.9 metres.

'But the tragedy goes deeper than that,' I tell the class. For what was he actually up to, rooting around in the soil like that? The priest brought to the surface facts that undermined the doctrines of the church. Didn't the discovery of fossilised pygmy elephants run counter to the creation of the world as described in Genesis? How did he explain this to his Indonesian seminary students? Forget the first seven days, forget the forbidden fruit, the insinuating serpent, the expulsion from the Garden of Eden. In the beginning there were stegodons...

In addition to his birth and death dates (1907–1990), several internet search hits included the abbreviation 'SVD' after his name. This, it turned out, stood for *Societas Verbi Divini*, or the Society of the Divine Word, a Catholic congregation founded in the Limburg town of Steyl, along the Meuse River close to Tegelen.

In 1948, at the age of forty-one, Verhoeven was sent out from the institute at Steyl to serve as a missionary in Flores; he boarded the SS *Kaloeloe* at Amsterdam for a six-week journey. Did he feel drawn by a sense of adventure, or did he, like the poet Jan Slauerhoff, suffer from *Herausweh*, or 'let's-get-out-of-here sickness'? Was there an invisible hand herding him along the gangplank of that ocean liner, away from home? I talk about the push and pull factors that play a role in every migration. Many missionaries were refugees from the poverty and dead-end prospects of their parental homes, seeking a better life in a seminary.

Did Theodor Verhoeven have children? He was a priest and had taken a vow of celibacy, but Verhoeven may have had children out of wedlock. We don't know that, I tell the class, but we also can't rule it out beforehand.

'Well, he *was* married,' someone says.

From the corner of my eye I had seen her pull out her smart-phone. Should I say something about it, I wondered. I can't imagine that is allowed in class. Just as I am about to crack the whip, she raises her hand. Unlike what I had supposed, she has been searching the web right here and now, looking for information about Theodor Verhoeven. In a specialised dictionary of mammal names, she has found something in the entry titled 'Verhoeven's Giant Tree Rat'.

'Would you like me to read it out loud?' she asks.

Father Dr Theodor Verhoeven was a Dutch archaeologist who was also a Catholic missionary in Indonesia. After twenty years as a priest . . . he left the priesthood, married his secretary, and returned to Europe.

2

BACK IN THE NETHERLANDS, right behind the station at Tegelen, a rocket-shaped minaret pokes at the sky. We have come here for a monastery, and we get a mosque. Our meeting point at Tegelen lies atop the highest terrace along the Meuse, at the edge of gullies of the dense riverine clay that the Romans once used to bake their *tegulae*, or roof tiles.

As it turns out, I'm the advance guard. To keep from wasting any time later on, I stop to get my bearings on the square in front of the station and check on Google Maps the route we will take on foot to Steyl. Parallel to the tracks, huge bites have been taken from the landscape: they are clay quarries, one of which proves to still be operative, for the manufacturing of roofing tiles.

I try to train myself in this: in 'reading' landscapes. Scratches, scars, gouges – who inflicted them? When one succeeds in looking through the lens of time, unexpected panoramas present themselves. It's what geologists do all the time: to every rock wall or sedimentary deposit they add an extra dimension, that of 'deep time'. The notion that the clay at Tegelen was formed two million years ago turns the holes in the landscape into prehistoric chasms: at the bottom of the quarries, fossils have been found of rhinos and hippos, panthers and apes, swamp turtles and porcupines, the large deer of Tegelen and the small deer of Tegelen. Right beneath my feet lies an assortment of fauna as fantastical as that of Liang Bua on Flores.

But we're here today for a different kind of digging. Once everyone has arrived and we've all left the station, I pass around the attendance list. There are fifteen of us. Downhill, in the monastic community of Steyl, across from the ramp for the ferry across the Meuse, a tour guide from the Society of the Divine Word is waiting for us.

We walk down Steylerstraat in little groups, until we come to a brick wall surrounding a botanical garden with three cloisters. Before we reach the gate, teaching assistant Pien, a junior in Dutch and a violinist, comes up with an idea. She's thinking about starting her article with the burial of Father Verhoeven, in 1990.

'Ashes to ashes, as it were,' she says.

It surprises me. A funeral as the start of a story. In my mind's eye I see an opening scene that mirrors the excavation of Flores Man, thirteen years later. Then, like an old-fashioned set of scales that, suddenly, unexpectedly tips to one side: the lowering of the coffin containing the remains of Father Verhoeven and the rising again of *Homo floresiensis*. Talk about visual sequencing.

The only thing is that I doubt whether Theodor Verhoeven is actually buried here in the graveyard at Steyl. A man who turned his back on the priesthood and married – isn't that reason enough for excommunication? But maybe I've missed something; perhaps the Catholic Church has kept up with the times.

*   *   *

JUST SO THAT we feel we've truly arrived, we cross the Meuse on the Steyl–Baarlo ferry, then cross back again. When we don't disembark, the ferryman only charges us for a one-way ticket. He lets two inland waterway vessels pass before starting back; everything seems to move slowly here. Standing on the forward deck, heading towards the ponderous chapels and cloisters of Steyl, we make our staged entrance.

But we are not blank slates. We've put in a lot of work in the last few weeks, cracking the background stories. As a detective squad, our breakthrough came by way of a lead that would make a thriller writer's mouth water: one of the chunkiest species of rat excavated by Father Verhoeven was named after a certain 'Paula'. In 1980, the rodent specialist at the American Museum of Natural History identified this specimen as belonging to a separate genus. It therefore required a name of its own, *Paulamys naso*: 'Paula's nose rat'. The creature's taxonomic description included two special details:

- Although originally thought to be extinct, *Paulamys naso* was spotted in the forests of West Flores in 1989.
- It was named after the wife of Dr Theodor Verhoeven.

Student Mariëlle has uncovered Paula's maiden name, as well as a scan of her obituary card in an online picture gallery. Paula Hamerlinck was born in 1904 at Evergem, and died in 2001 in Eeklo, both of which are in Flanders. 'Life partner of the late Theo Verhoeven,' it reads.

'Paula Hamerlinck was a former nun,' Mariëlle notes in our Dropbox. 'It's not clear how they met. In any case, they were both well along in years and never had children.'

Although Mariëlle is not the youngest in the group (she worked for fifteen years as a government employee), she has the best online research skills. Hour after hour she spends in digital mineshafts like Delpher and Dekrantvantoen.nl, going deeper than anyone else.

In one fluid motion she has also dug up a century-old application by Petrus Verhoeven, the priest's father, to build a bakery at Uden, on the high fens of Brabant province that drain into the Meuse. But then things went awry: his wife died the next year, in 1917, after the birth of their tenth child. Theodor was number five.

In 1990, Theodor Verhoeven's own obituary card stated:

He was ten years old when his mother died. From those earliest days, Theo began walking to the mission house on his own and ringing the bell. Was there any room for him yet? Week after week, until he was taken in.

In this document, Paula is referred to as 'his great love from the autumn of his life'. Together, they had gone through moments of existential doubt:

He was a wise man who had remained childlike, and so he loved children. Children who do not yet know our difficulties with the mystery of God, life and dying.

His greatest accomplishment is noted with pride: the international headlines about his discovery of fossils on Flores. 'He looked past the earth at the cosmos, at what reached beyond.' That, I feel, was well put, especially when one considers that although Theodor Verhoeven was not an astronomer, he had been in search of vistas too as he peered down into earthly pits.

His thesis on Church Father Tertullian's concept of the Trinity is in the university library at Leiden. It had made it past the ecclesiastical censor, and contains comments like: 'It is not commonplace to distance oneself from commonplace, ordinary ideas.'

The catalogue included a reference to *H 1429: Collection of Dr Theodorus Verhoeven SVD (1907–1990), with notes, articles, offprints and maps of archaeological excavations. From 1950, 1½ boxes.* This collection can be found in the attic of the Special Collections archive. Four of us go to work on it.

The special thing about the Special Collections archive is that before you are handed a box of archives, it is first weighed, down to the hundredth of a gram. When you turn it in, it is placed on the same set of electronic scales – a ritual every visitor undergoes. We picture an Indonesian cultural anthropologist using this fact to measure how much trust the Dutch place in each other (result: barely a hundredth of a gram).

The Verhoeven collection contains no brittle, antiquated documents, but it does contain a few (semi-)scientific publications with photos of bone fragments and digs. In one of them, we see Theodor Verhoeven himself. He is bent over a shallow pit, leaning with his right hand against an earthen wall; with his left he is making a digging motion. Unlike the startled Indonesian boy in the background, he seems earnest, inward-looking. His hair is neatly combed; he wears it parted on one side, short above the ears. His posture and body language express the dedication of – and I can find no other word for it – a musician.

Then there are the drawings. These are maps of Flores's coastline, a through road winding along the island's median, the courses of its rivers, and the caves. 'Liang Mommer' turns out to have been named after Father Mommersteeg, 'Liang Bekkum' after Vicar Van Bekkum. From another folder emerges one page of Verhoeven's journal – perhaps the most important of all. The date in the top corner reads 28 August 1950.

> *We take the car out early in the morning, Father Mommersteeg, Father Piet Smits [and I], from Ruteng to Téras. One of these [local caves] is very big. For a few years, it served as schoolroom for a number of classes at the same time.*

The village children who accompany them each carry a *tofa*, and a drawing in the margin shows a sort of hoe with an iron tip and a handle 'some 60 centimetres long'. Father Verhoeven is impressed by the 'beautiful stalactite, which is several metres in length'.

> *At the deep end of the cave we dig a pit, 1½ metres long and 75 centimetres wide. The top 20 centimetres contain a great deal of flint. The tofas are not much use, and we have no shovel. It is a good lesson. From now on, take a few samples first before starting a dig. The cave is called Liang (= cave) Bua (= cold).*

This reminds me of the game Battleship. On a grid of ten by ten squares, one hides an aircraft carrier, a torpedo boat, a minesweeper. You and your opponent are out to sink each other's fleet. Taking turns, you drop a bomb on one of the target squares – C7 or F2, for instance. Your hopes are fixed on the submarine, which is only two squares long and therefore the hardest to locate. On an August day in 1950, three Dutch missionaries showed up at Liang Bua on Flores. They selected a square and subjected it to the ruffle of the local children's *tofas* – right above the skull of *Homo floresiensis*, which would be heralded half a century later in the esteemed journal *Science* as 'the most sensational discovery in any branch of science for a decade'.

In March of 1952, Father Verhoeven returned to Liang Bua with his own seminary students, armed with spades this time. Three pits, each at a carefully measured spot, turned up layers of ash, as well as a ceramic pot. 'Thirty centimetres down we found bones,' he noted. In a slightly pitying tone, he described how some of his diggers fled when they encountered a human hand – for fear of evil spirits.

In the summer of 1954, in the Liang Toge cave, Father Verhoeven unearthed – amid the remains of a giant bat – an entire human skeleton. The skull was broken; he went to great pains to free it from the earth with a brush. The missionary packaged the bones in communion tins and sent them by packet boat to Surabaya, on Java, and from there to the Netherlands. One of the tins was sent by accident to the address of Verhoeven's sister in the town of Uden.

From the correspondence with a physical anthropologist at Utrecht University: 'These skull fragments are a fascinating find. We are rapt with attention.'

No lesser expert than Holland's most famous fossil hunter, Professor Dr G.H.R. von Koenigswald, announced the test results: 'To the great joy of all involved, this skull shows itself to have belonged to a Negrito.' It concerned a 'member of the long-skulled Asiatic pygmies': a modern human, *Homo sapiens*, but then a very early specimen: a proto-Negrito. In life, the inhabitant of Liang Toge had measured 1.46 metres.

As the finder and an amateur archaeologist, Father Verhoeven set out on publications of his own – in the Swiss ecclesiastical journal *Anthropos*: 'The skeleton has many archaic features,' he wrote in

German. The pelvic structure, the sharpness of the incisors and the shape of the skull caused him to speculate aloud about a possible descent from a local, as-yet-unknown form of primitive human.

Verhoeven continued digging. In early 1957 he found the jaw of an elephant, this time not in a cave but along the steep banks of a dry riverbed, a landscape element familiar to him from the Meuse Valley. The story first made it into the local paper, the *Javabode*, and then into Verhoeven's native daily paper, the *Maasbode*. 'Fossil of Antediluvian Creature Discovered on Flores,' the headline read.

The section of jaw, with upper and lower teeth, belonged to a pygmy elephant that lived half a million years ago. 'What a wonderful discovery, deserving of heartfelt congratulations,' Von Koenigswald wrote. 'I had never thought that our elephants had made their way so far to the east.'

The finder himself added here, not without a certain pride, that Alfred Russel Wallace ('a friend of Darwin's') had noted in 1859 that the great Asiatic mammals had come no further than Bali. That is to say: never past the strait between Bali and Lombok. 'This used to be called the Wallace Line.' But now the missionary Verhoeven from Uden had found the fossils of elephants on Flores, two islands to the east of that. The British press called it 'a heavy blow' to the Wallace Line.

Nowhere in his notes does Theodor Verhoeven show himself to be vain or in any other way ill-tempered. It remained unclear whether Paula went through the items in the archive boxes beforehand to weed out any unseemliness. Amid the correspondence concerning their acquisition by the university, we found a letter in her hand-writing. On his deathbed, she wrote, her husband had stated his wish to leave his private collection not to the archives of a cloister, but to a school.

'He was already very ill at that time [April 1990],' she noted. 'He died on 3 June 1990.'

BY THE TIME we arrive at the monastic community of Steyl, we've noticed one thing for certain: the lack of a single expression of religiosity or spirituality at any point in the life of Father Theodor Verhoeven. Yes, he attended both the minor and major seminaries, and he had been ordained and sent out as a missionary to Flores to proclaim God's word. And we know how that ended up: around Christmas of 1966, Verhoeven lost control of the wheel of his mission vehicle. The priest missed the curve and plunged into a ravine. He broke so many bones that he had to be repatriated to the Netherlands. 'I had to go back to Europe immediately,' was his businesslike take on the matter. The accident marked the end of his eighteen-year stay on Flores.

MISSION ACCOMPLISHED?!
It seems, as one approaches Steyl aboard the Maashopper – the local ferryboat – that a medieval village is rising up before your eyes. The citadel you think you saw, however, is the twin-spired mission church of Steyl.

That is how Elizabeth, a teaching student from Zeeland province, begins her article. Perhaps she had in the back of her mind the Great Flood of 1953 in her own home province, when she focuses in the next few lines on the three memorial plaques in the church wall at Steyl, showing the high-water marks from the twentieth century: 1926 (when the altar was found floating through the church), 1993 and 1995 (when the damage consisted only of wet floorboards and kneelers).

The Meuse is capable of swelling and leaving its banks within a matter of hours, and then, in summer and autumn, sinking back into lethargy.

THE DENIM-CLAD VOLUNTEER who is waiting for us on the square across from the ferry ramp speaks in the velvety accent of the far Dutch south. His name, he says, is Karel. I assume he is a retired priest – there are about thirty of them here. These days, the brick-clad St Michaëlmannen cloister rising up behind us serves as an old folks' home.

'At its peak, before the Second World War, there were seven hundred priests living here,' Karel says. Striding through the monastery garden, he leads us to the catacombs the monks had dug into a hillside: subterranean corridors with statues of the saints in dimly lit niches, as organ music babbles from the loudspeakers.

Around the hill itself lies a boiler house, two nunneries, a print-shop, a studio for the manufacture of holy figures, as well as the country's first heated horticultural greenhouse, where vegetables were grown for the friars and nuns.

Stadt Gottes was its German name, meaning City of God. The order was founded in 1875 by a priest from Germany, Arnold Janssen. His portrait is omnipresent. The round-headed abbot had been canonised by the Vatican, and was therefore a fitting subject for public veneration. But where are the living, breathing, singing clerics? The students notice their absence too; most of them will mention it in their articles.

Is Steyl a ghost town?

Have the friars and nuns gone into hiding?

Was it too cold for them?

But they are there. Karel points out to us the two nunneries: one for the blue sisters, the other for the pink, according to the colour of their habits. The blue nuns are allowed outside; they are mission sisters, mostly nurses. The pink nuns are sequestered in an enclosed convent and spend their days in prayer. The Servants of Perpetual Adoration, they call themselves.

'They have closed the door to the world behind them,' Karel tells us. 'Even when they die, they don't go outside; their body is taken directly to the cemetery through an underground tunnel.'

One door to the chapel of the pink nuns is always open: any layperson can go in, day or night, seven days a week, to see a nun at prayer, kneeling there behind cast-iron bars. The place is the religious heart and soul of Steyl.

As I climb the steps to the public gallery, I cannot help but feel that I'm going to observe some exotic, endangered species of human. Would I view the caged nuns in the same way I might the flamingos or the ibises at the Amsterdam Zoo?

In his thesis, Father Verhoeven warned against the dangers of complacency. He provided a quote in French that came down to this: for those unfamiliar with them, religious expressions are easy to laugh off.

The chapel is filled with women's voices. I recognise the 'Our Father'. All the Servants of Perpetual Adoration have gathered for their daily Mass and are sitting in the pews, heads bowed. Parked in the aisle is a wheelie-walker. At the unanimous 'amen', the nuns rise and file out a door at the back of the chapel. All except one. That nun walks up to the prie-dieu before the altar. Kneeling, she folds her hands and lifts her head to face the star-shaped holder containing the consecrated Host: a piece of Christ's body. She will remain in that position for the next hour, until, in a changing of the guards, another sister takes her place.

It is to the sisters' credit that they do not care what I think. 'We are in the world, but not of the world,' their brochure says. For them there is no marriage, no children, no family. The sisters step knowingly and willingly away from the norm, and it's no little step at that. They pray unceasingly on behalf of 'our brothers and sisters from Steyl, who work as missionaries around the globe'. As the brides of Christ, they recommend to God priests like Theodor Verhoeven: that he not be brought into temptation in faraway places. 'We firmly believe that our prayers bear fruit, otherwise our presence here would have no purpose.'

But still, no matter how much soul power the Servants of Perpetual Adoration brought to bear, Theodor Verhoeven and Paula Hamerlinck had one day taken each other's hand. Together they had jumped ship.

OUTSIDE, IN THE alleyways of Steyl, Karel tells us that the perpetual adoration will soon be coming to an end. He has based his prediction on simple figures: around 1900, the number of pink nuns had peaked at more than two hundred, but natural attrition and a lack of new novices mean there are now only sixteen left. 'They won't be able to keep this up much longer,' our guide says flatly.

Suddenly it occurs to me that Karel might not be a retired missionary after all.

'I'm a biologist,' he says. 'I've renounced the Church.'

The order in which he says this intrigues me, as though the second is a mathematical product of the first. Might a similar realisation have overcome Theodor Verhoeven too?

Our guide knows a great deal about Steyl, but nothing about Verhoeven. In the Mission Museum in the central square is a register of the names of all the priests of Steyl, living and dead. Karel leads us to the reception desk and begins leafing through it. He gives up quickly enough. 'Verhoeven, Theodorus Lambertus' is hard to find, and he has another appointment that afternoon.

After saying our goodbyes, we turn back to the book, which has a green hardbound cover, but no matter how we search – by name or date of birth – it appears that Father Verhoeven, unlike Father Mommersteeg or Vicar Van Bekkum, has been scraped relentlessly from the memory of the Society of the Divine Word.

The only other spot where Verhoeven's shade might still walk these halls is in the Mission Museum itself. Entertaining no great expectations, we shuffle across the parquetry floor to the exhibits. We pass a stuffed bear, Russian, upright, moth-eaten. When you toss a coin into the slot near its heart, the animal is meant to growl threateningly for you – but the mechanism running its internal bellows has apparently worn out after eighty-five years, and now it only groans.

After passing a bust of Father Arnold Janssen, we enter the insect room.

'Until I saw it with my own eyes, I had never known that butterflies could be so blue,' Elizabeth will write in her article.

Like children, we mill around the display cases, which are crawling with creatures. Lemmings, marmots, meerkats. It is the sheer number of prepared animals that impresses the group, as well as the care with which they have been displayed, even in a scene showing a lioness biting a zebra's throat.

'Hairy dwarf porcupine, pinnated grouse and bearded vulture,' was Elizabeth's summary. I myself noted: 'Goliath beetle, boa constrictor, human skulls from the head-hunters of Papua New Guinea.' More than a natural history cabinet, this is a shrine. The pinned butterflies are arranged by colour: blue, bluer, bluest. Animal beauty to the honour and glory of Creation.

'The priests from Steyl brought local fauna from all over the world back to Steyl,' Bob will write. 'In this way, a collection was assembled to chart God's creation, with Man at its pinnacle.'

Meanwhile, we continue our search for traces of Theodor Verhoeven. We find them at the last moment, in the form of two monitor lizards from Flores. Komodo dragons, a male and a female. The absolute eye-catchers of one of the dioramas, they stand there, just being gigantic.

3

A DOUBLE PORTRAIT, A DUTCH GIRL with an Indonesian man, pops up on my monitor. They pose side by side beneath the rocky overhang at Liang Bua; the cave entrance itself is out of sight. The girl is fifteen, not yet fully grown, yet she towers over the adult male beside her. He is a rice farmer who showed up specially for the photo. The little man is barefoot, but that does not explain the difference in height: the girl is plainly two heads taller.

I had written to her mother, José Joordens, asking if she had any hominid skulls for us. At the top of my wish list was that of LB1, but material by way of comparison would be welcome too. Dr Joordens is a member of the Human Origins Group in Leiden. Three simple words. Human. Origins. Group.

The tone of José's reply is easy and familiar: 'Here's a photograph of my daughter Julie with one of the legendary little people who still live around Liang Bua. Flores is truly a lost world. Fascinating.'

Along with my students, I'm hoping to make the leap from the abstract to the concrete – with the aid of skulls one can hold in one's hands. Solid matter. Reportage is not a mollusc; the genre belongs in the category of vertebrate stories that take their strength from a backbone of fact.

José tells me she can arrange for casts of all the Flores fossils. 'Maybe a few weird bones from Father Verhoeven's island fauna too?'

It is all being handed to me on a silver platter. Will I even need to go to Flores later on?

The Human Origins Group has its offices about one kilometre from the Faculty of Humanities. On their website, the fourteen staff members are seen sitting in a circle; in their midst is a shop-window mannequin in the guise of a hairy primordial man.

Two disciplines merge, I read, within the Human Origins Group. Archaeology on the one hand, and on the other that branch of human biology that reaches back beyond what is recorded in books or scrolls of papyrus: paleoanthropology. 'Paleo' comes from the Greek *palaios*, 'old'. Anyone wishing to leaf back through human history further than what is found written on clay tablets must go digging in the clay itself. Then you automatically arrive in prehistory.

The earth's layers, fortunately, allow themselves to be read like the pages of a book. Some of those layers can be dated quite accurately, as though marked with a page number. Almost the entire substrate of Denmark, for example, is marked by a two- to three-millimetre-thick layer of clay containing a high concentration of iridium: a sort of bookmark telling the story of a volcanic eruption on the Yucatán Peninsula in Mexico, some 65.9 million years ago.

In this underground archive, the paleoanthropologist searches for anthropoid fossils, the remains of hominids. Skulls and dental fragments provide the best information, especially in combination with the age of the substrate where they are found. Then the archaeologists come into play. They dig for stone axes, spearheads and the remains of hearths – artefacts that shed light on their human makers.

The objective that the Human Origins Group has set for itself is to collate the skeletons of hominids with the tools found, in an attempt to determine what has made humans human. In what way does *Homo sapiens* differ from other mammals? Or, to paraphrase: who are we?

José says she'll need a few days to collect the things we need. First she has to talk to an Indonesian TV crew about the Dubois Collection. She needs to prepare what to say and what not: the subject is a tricky one. The Dubois Collection is Holland's biggest fossil treasure-trove of colonial spoils, shipped from Java in four hundred crates in the late nineteenth century. All told, it contains some 40,000 bones, teeth and shells from Java and Sumatra, unearthed between 1888 and 1895 under the supervision of Eugène Dubois, the son of a pharmacist from Eijsden, a village south of Maastricht where the Meuse River enters the Netherlands.

The collection's former storage space in Leiden used to be marked with a sign reading 'INDONESIAN FOSSILS (DUBOIS)' – followed by the admonition: 'CLOSED TO THE GENERAL PUBLIC'. These days, all the relics have been moved to a specially built 'collection tower' at the Naturalis Museum, a 62-metre-high vault in which Professor Dubois' trophies take up an entire floor.

The absolute showpiece is kept in a fireproof metal safe and consists of three parts: a femur, a molar and the top of a skull. These belong to 'Java Man', the first unearthed (in 1891 and 1892) remains of our direct ancestor, *Homo erectus*. The fame of its finder Dubois (1858–1940), after whom a planetoid has since been named, is based on this three-in-one of skull plate, femur and molar, heralded at the time as the relics of the 'missing link' between humans and animals.

'Eugène Dubois is my scientific hero,' José Joordens writes. (Later she tells me that she occasionally places a candle on his grave at Venlo.)

I'm too late to visit the tower itself: the entire collection has been packed in dust-free containers for an upcoming renovation. 'But I can bring you a cast of an African *Homo erectus*,' José tells me. My students and I are welcome to meet her at Einsteinweg 2, in the office of a professor who is currently on sabbatical.

José has also invited the former curator of the Dubois Collection: the man who spent the lion's share of his professional career watching over these fossilised crown jewels. John de Vos is retired now, but suffers from withdrawal symptoms, so he is all too pleased to come to Leiden to see us. 'John is a specialist in human evolution,' José says. 'He'll bring one of the little skulls of *Homo floresiensis* along with him.'

MY ENTHUSIASM IS growing, but I'm having a hard time passing it along to my students. 'It just doesn't really interest me,' says Els, a 55-year-old art historian. For five long weeks now I've been beating her about the ears with pygmy elephants and giant rats. 'You know what it is? I'm a Catholic. When you start in about ape-men, I think: *Yeah, yeah. I happen to believe in Creation.*'

This is a harder blow than I care to admit. Els is the only student who is older than me. I consider saying: 'You're not in church here, Els, you're at the university.' But instead I look around the circle of faces: what do the others think?

Lian accepts the challenge. 'If you ask me, who you are has to do partly with the question of where you come from.' Her Asian features – is she an adopted child? Were her parents Vietnamese boat people? – lend additional weight to her statement. 'The answer to that question, I think, is just as important for a person as it is for a species.'

After class, Els comes to me to ask if she can take my picture. She paints portraits, and she'd like to make one of me. I consent without enthusiasm, which serves to increase the uneasiness between us. Looking into the lens, I ask myself why it is that I am so fascinated by the skulls of hominids.

The point Lian made about origins as one of the keys to who we are seems to me to hit the nail on the head. Long ago, the genus *Homo* split off from the apes. Skulls have been found of our ancestors, the earliest hominids. Now I want to hold one in my hands. Why? I want to lay my finger on that one groove or bulge that makes our species unique. On the characteristic that makes tangible the difference with the chimpanzee or orangutan. If one accepts the story of Adam and Eve for what it is (a story), then what is there left to say about the origins of humanity?

If we imagine a 'cosmic calendar' which equates the age of the universe with one year, then the Big Bang took place at 12 a.m. on 1 January, and *Homo sapiens* appeared on New Year's Eve at one minute to midnight. The earliest hominids appeared at 10.30 p.m. At fourteen minutes before twelve, our ancestors succeeded in controlling fire. ('Just in time to set off the fireworks,' says history student Roger.) Our species is the only one with the destructive power to – with one push of the nuclear button – extinguish all life on earth. Admittedly, I think people are special. Not more subtle than other creatures; no, not that. But definitely unique. Extravagant.

To the primatologist who deduces from chimpanzee or bonobo behaviour that humans and animals are not fundamentally different,

I want to shout: 'Still, chimpanzees have no incubators for their pre-maturely born, no spy satellites, no deep-freeze dinners, no cigarettes or warning stickers reading SMOKING KILLS, no mortgages or indi-vidual retirement funds, no Tomahawk missiles, no tax deductions for registered charities, no criminal attorneys, no symphony orches-tras, no torture chambers, no CT scanners, no confessionals or prayer rugs, no casinos, no brothels, no seeing-eye dogs.' We humans are the only creatures in the world who tell each other stories; we can blush and tell jokes, program computers and commit acts of sabotage, phi-losophise and play-act.

This is not an attempt on my part to congratulate the species on its uniqueness. Only to position it in a category of its own. For the purposes of observation. But how am I supposed to get Els onboard?

My other students are hobbled by a different sentiment: like hams in a smokehouse, they are permeated with postmodernism; for them, reality exists at most as a wispy cobweb of playful references. Nothing 'really' exists – the equator, a semester, north and south; they're all arbitrary agreements. Elfrieda, who is studying Dutch in her junior year, responded to my first lecture with the words: 'Every claim you make leaves me thinking: no, no, no. That's not the way it is!'

Most of them have developed feelers of suspicion concerning hard, reliable facts – there is no such thing. Figures? Measurements? Radiocarbon dating?

'Our drive to measure everything and everyone – what does that say about us?' That was the contribution of the professor of modern Dutch literature who often sat in on our sessions. She dragged in a dead French philosopher; I came up with a BBC report about a study at the Imperial College in London: 'Dutch Men Revealed as World's Tallest'.

Anthropometry may have been contaminated by colonial abuses, yet in 2016 Dutch men turned out to be the planet's tallest. With an average height of 183 centimetres, they are at the top of the international list. The discrepancy with the average East Timorese male, at barely 160 centimetres, is 23 centimetres and growing.

The professor who was instrumental in getting me my residency at Leiden asked my students: 'Why would we want to classify each other? Where does that urge come from?'

Grr. First the facts, then the questions. Dutch women are tall as well (169 centimetres), but still stand second to Latvian females (who have an average height of 170 centimetres).

The person wielding the ruler is, of course, in a different position from the one being measured. The question 'Who is measuring whom?' is a legitimate one. But does that invalidate the measurements themselves? Frank Zappa, I read somewhere, once said: 'Without deviance from normality, there can be no progress.' Agreed, but then why not first figure out which norm you wish to deviate from?

The trend between 1914 and 2014 – the century held up to the light by the Imperial College researchers – has been: people grow taller, but not all of them, and the growth spurt is not the same everywhere. The variation has to do with differences in welfare, health care and nutrition. Did dairy products account for the Dutch record? The consumption of cheese? One researcher saw a connection between democracy and increased height – a dictatorship is out to keep its subjects down. Pride helps you grow, bowing makes you shrink.

Homo sapiens is not finished yet. For those who see modern humans as the end product of evolution, I come waltzing in with the stature of the *Übermensch* – who, according to his creator, is already looking down upon us from future heights 'in painful embarrassment' at our sorry figures.

Nietzsche got the students' attention, but I was planning to stick to the facts for the time being. Shortly before our visit to José Joordens and John de Vos, the Faculty of Theoretical Biology at the University of Vienna released news that appealed to my imagination: 'Caesarean Sections Affect Human Evolution'. Medical statistics had revealed a pattern: the average 'Western female's' birth canal had become measurably narrower in recent decades. Thanks to the Caesarean section, a woman with a narrow birth canal no longer dies along with her baby at childbirth, and so is able to pass along this genetic trait. The gist of this discovery is that by intervening surgically in the birth process, humans are modifying their anatomy. What other animal species tinkers with the mechanics of its own evolutionary process?

José Joordens is breaking bars of fair-trade chocolate into pieces. She removes honey waffles from a package and puts them on plates scattered around the table where our group has gathered. With a barely audible zoom, a beamer projects on the wall a palm-lined beach as landscape decor.

José and I first met at a New Year's party in Amsterdam ten years ago. Her daughter and mine were then six and ten. We both remember the fog that hid the beautiful glare of the fireworks from view.

Since then, José has caused a furore with her scoop in *Nature*, in December 2014. At one fell swoop she made a name for herself with a discovery from the 125-year-old Dubois Collection. Dr J. Joordens was the 'first author' of a startling publication dealing with one of Dubois' primordial mussel shells, which turned out to have lines etched on it, in a sort of drawing. Along with twenty-one co-authors, José succeeded in proving that that shape had been intentionally scratched on the shell. It is a zigzag line, put there by an adroit hand. The lines on the shell form a pattern, but a pattern that is half a million years old. *Homo sapiens* had not yet come on the scene, which meant the drawer must have been its predecessor, *Homo erectus*. Who was unearthed by Eugène Dubois from that same high, clayey bank of the Solo River on Java, where the shells were found too.

In other words: the trio of skullcap, femur and molar belonged to a primitive hominid who, it had now been shown, was *able to draw*. It was as though José had administered the Javanese ape-man with an extra dose of 'humanness': in addition to brutishness, therefore, *Homo erectus* also had a more refined, perhaps even artistic side.

'How many of you are coming, again?' José asks, pondering the seating arrangement. She wants me to sit across from John de Vos. I'm expecting at least twelve students, but there could be twenty. Just to be sure, José summons her teaching assistant: can he arrange to get a few more chairs?

Once the preparations are over, I tell her that I have booked a flight to Flores in late April, for myself and for Vera, who will be fifteen by then. And also: that I showed my daughter the photo of Julie and the little Flores man at the entrance to Liang Bua. 'Oh, but then I won't be short over there,' was her reaction.

José predicts that we will meet that man too: they drum him up every time, as a living attraction, whether you ask for it or not.

'Yohanes Dak,' I nod.

That's news to José; she had given him a few rupiahs, but had not asked his name.

Yohanes Dak cashes in on being small. One of the students recognised him in the photo from Liang Bua; he was, as it turned out,
the very same Yohanes Dak who had graced the front page of the
Jakarta Post one day in 2005. In the accompanying article, Professor
Teuku Jacob, Indonesia's number one expert on fossils, pointed him
out as a living example of *Homo floresiensis*. The distinguished Jacob,
an authority on the discovery of archaic human remains, used him
to ridicule all the hubbub about LB1. Flores Man was a hoax. People
on Flores just happen to be little, 75-year-old Jacob said. Just look at
Mr Dak! The rice farmer of Liang Bua stood ramrod-straight under
the stadiometer, as though hopping to attention. In the black-and-
white photo in the *Jakarta Post*, he is no taller than 125 centimetres.

José's teaching assistant comes in with hatboxes. From the tissue-
paper lining he produces skulls. Big, whole ones, yellowish in colour.
But also brown ones with cracks and holes in them. Jawbones that
can be separated from the top section of the skull, the cranium. As
well as loose, bowl-shaped brainpans. He puts them down among the
plates of cookies and chocolate.

THE OPENING LINE of Bob's article will read: 'Lying on the table
is three million years of human history, reduced to a dozen skulls.'

WE'VE HAD OUR eye on Teuku Jacob for weeks now. He is an
éminence grise and an *enfant terrible* rolled into one. Fiercely national-
istic, Acehnese, born at the northern tip of Sumatra on 6 December
1929. Because of his age, he was able to squeak through World War II
as a boy. During the chaotic post-war years after Hiroshima and

Nagasaki, the capitulation and the Japanese retreat, he worked on
patriotic radio broadcasts. His own major battle was one he fought
later: during his career, against the Dutch colonisers, already driven
out in 1949.

To that end, Jacob could not do without a doctoral degree from
a Dutch university, which he took under the supervision of *tuan*
Von Koenigswald at Utrecht. *Tuan* is Malay for 'master'. To initiate
his Indonesian pupil into paleoanthropology, the study of the fossil
remains of ancient and even more ancient hominids, Von Koenigswald
gave him access to the skeleton of the 'proto-Negrito' from Flores,
Father Verhoeven's discovery.

> We are only scattered bones
> but they are yours
> you have to decide the value of those scattered bones.

His thesis, completed in 1967, bore this motto from an Indonesian
poet.

On the first page, Teuku Jacob – who was thirty-eight at the
time – holds up 'the person of Dubois' as the unrivalled pioneer of
paleoanthropology, someone who 'towered above' those who followed. 'With insight, dedication and good fortune, he found the first
specimen of an ape-man.'

His supervisor, Dr Ralph von Koenigswald, then received praise
for his discovery of more archaic hominid remains between 1936 and
1941. Teuku Jacob regarded both Dutchmen as the founding fathers
of his chosen field, but deeply disapproved of how they had exploited
their Javanese personnel: in the late nineteenth century, Dubois had
used forced labourers, while Von Koenigswald paid his workers

starvation wages. The first discoveries on Java of the prehistoric skulls of ancestors of *Homo sapiens* had taken place in a truly colonial ambiance, with as its pioneers two Dutchmen who were blind to volatile reality. 'The war in the Pacific brought the research to a halt, just as all wars in the history of hominids have done, with regard to so much,' Jacob wrote.

Teuku Jacob's dissertation is factual and philosophical at the same time. We didn't have to go looking for a copy: there was one in Father Verhoeven's archive boxes, which was striking, in view of the fact that in it the doctoral student knocks the props out from under the missionary's work on one crucial point. According to Jacob, the Liang Toge skeleton is not particularly significant. He dates the bones shipped in those communion wafer boxes to be only 3500 years old, 4000 at the very most. That is much too recent to provide any insight at all into human evolution. The proto-Negrito, about which Von Koenigswald was so enthusiastic in 1957, was no proto-Negrito at all, nor was it an Asiatic pygmy. 'The group it represents might be short, but not dwarf.'

This was the blow administered to his superiors by Teuku Jacob in his 1967 thesis: the skeleton was not unique, not anomalous; it was simply 'normally' small.

And then, in his twilight years, after LB1 had produced its shockwaves, Teuku Jacob showed up again and launched the biggest offensive of his career: LB1 was a false alarm, much ado about nothing. 'Flo', as Flores Man was referred to fondly, was a figment created by pseudo-scientific conmen, he argued. The editors at *Nature* had fallen for it, or were part of the complot. In real life, Flo was nothing but a normal *Homo sapiens* who had suffered from microcephaly, dwarfism of the skull.

Professor Jacob, himself just 157 centimetres tall, put both his own reputation and that of his country on the line. He was a scientist and a politician – a former member of parliament. As soon as Jacob was contradicted by the Australian scientists who had found and classified Flo, he confiscated the skull and other bones of LBS 1 through 9. He spirited them away to his own laboratory in Jogjakarta, out of the reach of 'those Australian sheriffs'. Western hominid-hunters had no clue about how to behave in 'a nation that is still young'.

Teuku Jacob ordered the cancellation of all further excavations at Liang Bua. Even more drastically, in 2005 the limestone cave was sealed off. A fence was placed at its entrance, decked out with rolls of barbed wire.

4

W̲E TAKE OUR SEATS across from John de Vos and José
Joordens like a trade delegation. Attentively, expect-
antly. The former curator of the Dubois Collection is
seated halfway down one side of the long table. José introduces him,
starting chronologically with his thesis on dwarf deer on Crete and
ending with his curatorship.

'I was about everything that was old, dead and vertebrate,' John
says. His grey hair is combed back straight, and he sports a fringe of
beard that reminds me of Lenin.

FROM BOB'S REPORT: 'In the space of two hours, John de Vos –
his cane leans against the wall – and José Joordens – a generation
younger – unfold their vision of man, and what skulls tell us about him.'

THE OPENING MOVE – explaining who we are and what we're
doing here – is up to me. I mention my role as writer-in-residence.

John nods; he's familiar with the residency at Leiden, and has been ever since 1985, the year Gerard Reve set the ball rolling – and wounded a member of the audience at the closing session with a broken wineglass. People still talk about that.

As the first writer-in-residence (in thirty-two years) who does not dream up his own stories, there's no way I can simply shrug off the onus of Reve's 'Law of Impracticability of the Real'. Reve considered reality to be implausible, too good to be true. But I prefer to give that a twist: lots of things are too true to be good; it is precisely the facts that contain the drama. This leads me to the point of my explanation: we're here, at this table, for tangible material, for things you can hold in your hands. While I'm rounding off my little speech with something about deviance and norm, I catch myself staring into dozens of empty eye sockets. The nasal cavities are striking too. Missing teeth and molars.

'All right, but what is normal and what is anomalous?' John de Vos leans his forearms on the table. He's wearing a sportscoat with leather patches at the elbows. 'In fact, I don't know anything about that. I couldn't tell you.'

What he does know about: everything to do with Eugène Dubois and his 40,000 fossils. José – who is looking askance at him through her fringe of black curls – has warned me already: if I let him take over the helm, this will turn into an exposé of the ups and down of Dubois as the godfather of paleoanthropology.

'In *The Descent of Man*, Darwin wrote that man lost his fur in the tropics,' John continues. 'Darwin wasn't thinking of Asia, though – he was thinking of Africa.' But, John adds immediately, he holds Wallace in just as high regard as he does Darwin, perhaps even higher. In any case, Wallace deserves more recognition.

When it comes to the origins of the human species, though, Darwin's reasoning was purer: a transitional form must once have walked the earth, a missing link between the great apes at one end of the evolutionary chain and us at the other. Darwin imagined that missing link to have been an ape-man, a creature in between apes and humans.

'Eugène Dubois, whose father was a pharmacist and the village mayor, flees Catholic Limburg and goes to the Dutch East Indies in 1887 with no other objective than to find that missing link,' John says. Then, with surprising speed, he gets down to business. 'First, Dubois finds a molar, then this . . .'

He slides the skullcap across the table to me. It is worn smooth as a baby's rattle and has a slight ridge above the brow.

José: 'The men who found it thought it was a turtle shell.'

'Dubois himself wasn't there at the time,' says John. 'Two Dutch corporals had been assigned to the digs. They were there to keep an eye on about twenty Javanese convicts who had to do all the digging.'

'And they kept dying,' José says. 'So they had to be replaced.'

'They're the ones who found this.'

I touch the brownish bowl, tap it. John looks around and points. He's looking for the skull of an ape and one of a *Homo sapiens*, but they're not there. José's teaching assistant gets up to fetch them. In the doorway, he runs into three latecomers. Roger, a history student, makes his apologies. 'We couldn't find Einsteinweg.' While we slide our chairs up to make room, one of his companions chimes in: 'We never come to this side of the tracks.'

There is chuckling. I don't get the joke, but the professor of literature fills me in: here, to the west of the railroad tracks, the streets are named after Einstein, Newton and Archimedes. This is the domain of

the hard sciences, the 'bioscience park' that also houses the Human
Origins Group. For a humanities student, this is *terra incognita*, a place
one cycles past at most on one's way to the athletics fields.

Then José's assistant comes back with the requested goods.

'Look, this is a chimp, and this here is *Homo sapiens*.' John slides
the skulls forward on the board, like pawns, meanwhile nodding with
his chin at the brown skullcap. 'And then Dubois comes up with
this. You folks want to know what's normal and what's anomalous?
Well, you tell me.'

In front of us lie three skulls, like samples of some rare commod-
ity. John wants me to inspect them. I pick them up one by one and
provide a running commentary: 'Dubois' skullcap isn't from an ape,
because it doesn't have a seam across it, front to back, like a chimp's.
But maybe it's also not a human skull, because the forehead is flatter,
except for this ridge.'

John doesn't look pleased, but not displeased either. 'That's what
you say,' he says. 'Dubois, by the way, thought exactly the same thing.
And then he found this . . .'

I was handed a longish bone, brown as well.

'A left femur,' José clarifies.

'Which was lying fifteen metres away, in the same riverbed,'
says John.

We ask them how old this bone is.

'Half a million years,' José says.

'A million years, just like the molar and the skullcap,' says John.

There is a shudder of disbelief on our side of the table. A discrepancy of half a million years? That sounds like flim-flammery with the
centuries, a hustle in millennia.

'Do I hear any more bids?' Elfrieda asks.

That's not the point, though, John and José assure us. The important thing is that Dubois' fossils are much older than the oldest human. *Homo sapiens* made his entrance only 180,000 years ago.

'What we have here,' John says, 'is a very early biped.'

'In other words, a species that already walked upright,' says José.

Dubois deduced the creature's posture from the femur. This long, straight bone is not found in chimpanzees or orangutans. John de Vos holds it up, while explaining that in shape and strength the bone is built to carry the full body weight of a sizeable land mammal, without the support of front legs.

In a nutshell: in the late nineteenth century, Eugène Dubois of Eijsden discovered on Java a transitional form that was neither ape nor human. The missing link. Resounding evidence for the heretical idea, unacceptable at the time, that humankind too is a product of evolution. In full, Dubois named his transitional form *Pithecanthropus erectus*, the upright-walking ape-man. Later rechristened as *Homo erectus*.

BACK TO THE table. Of the ten skulls on display, the smallest belongs to Flores Man. LB1's jaws are clenched. Her death's-head grin reveals a miniature set of teeth, brittle and incomplete. Flo's eye sockets are so large that there is almost no forehead left, making it look like she's staring at us in amazement. The feeling is mutual.

For the purposes of my lecture series, I had tried to buy a cast of her skull. Bone Clones, Inc. had one in its web shop. *Homo floresiensis skull, LB1, approx. 18,000 years old. Full-scale replica: 13.5 cm. high, 15.5 long and 11.0 wide. One of the most important discoveries in decades; H. floresiensis is the subject of heated debate.*

Price: $325.

But as soon as I loaded LB1 into my shopping cart, a pop-up appeared: *Availability: discontinued.*

Skulls Unlimited International also had Flo in its assortment. *H. floresiensis is a small hominid, no larger than a modern human child (1 metre). H. floresiensis lived in isolation with dwarf elephants and Komodo dragons. Diet: omnivore.* And below that: *This item is no longer available.*

LB1 replicas were no longer in stock. Sold out?

Or not? When I click on the nine-part 'Hominid Skull Set Deluxe', I see that Flo is in there. For that price I'll also receive 3.2-million-year-old 'Lucy', found in Ethiopia's Afar Depression; on the evening of her discovery, the transistor radio was playing the Beatles' 'Lucy in the Sky with Diamonds'. The web shop had a link to a teachers' manual, a sort of instruction leaflet that links the finder with the found. This provided one with pairs such as 'Donald Johanson & Lucy' and 'Raymond Dart & the Taung Child'.

Lucy and the Taung Child were too primitive and too old (more than two million years) to fall within the genus of *Homo.* They were australopithecines – literally meaning 'southern apes' – but because they stood at the bottom of humankind's family tree, they were part of the set anyway.

The deluxe set cost $2122. On top of that was the postage cost ($80), while I was also given the opportunity to add accessories to my cart: a carrying case ('both attractive and lightweight') or a premium carrying case ('with vinyl coating').

BUT NOW IT'S our turn to start puzzling. For the first time, I hold Flo in the palm of my hand. It's the completeness that surprises me, in combination with the size. My association is not with a coconut, nor a grapefruit either – more like a softball.

The skull of LB1 is not only much smaller than the brainpan of the Javanese ape-man, it is also rounder, more of a real sphere. I run my finger around the inside. Flo, as it turns out, has a bigger frontal cavity than most hominids, which provided more space for the frontal lobes. This oyster-shell smooth cavity might have provided enough room for a highly developed capacity for anticipation, multitasking and communication.

'Limited volume here is accompanied by a relatively large communication centre,' John explains. 'One supposes that, at least in some ways, Flores Man was smart.'

Paleoneurology would seem to be a discipline unto itself. Still, I can hardly imagine brain volume failing to correspond to cognitive capacity.

'400 cc,' Roger comments. 'That's the brain capacity of a chimp, isn't it?'

John admits that Flo is not simply on the petite side; she is extremely small.

'Smaller than Lucy.' José draws our attention to the remains of Lucy, of which only half of the lower jaw and a few fragments from above the ears actually look old (they are a dark brown); the rest consists of an ivory-coloured reconstruction of how the skull must once have looked in its entirety.

Some of us are now standing around her. We're allowed to arrange the skulls in chronological order, as we see fit. Which form emerged from which? José helps by placing Lucy at the very front: she is the progenitrix. The remaining skulls we now line up, from large to small, or according to their resemblance to *Homo sapiens* (who has a pronounced chin and a high forehead). But no matter how we shuffle them about, a skull with heavy brow ridges always ends up in the middle; this half-man looks as though he wandered the steppes with a pair of bone goggles on his head the whole time.

We're curious about who he is.

'That's the earliest *Homo erectus*,' José lectures. 'He's 1.6 million years old.'

And does he have a name?

'KNM ER 3733.'

KNM apparently stands for Kenya National Museum; ER for the site where he was found, East of Lake Rudolf, a safari park where the Anglo-Kenyan Leakey family (father, mother, sons, daughters-in-law, grandchildren) is wont to hunt after hominid remains, the way wealthy dentists pursue the 'Big Five' game animals.

LB1 remains an odd duck: here, on the table in front of us, her miniature skull won't let itself be fitted into the ancestral family tree.

'The more we find, the less we know.' José leans over the table, assuming the same pose as John, who has made a trademark out of provocation. 'With every new discovery, it just gets messier and messier. Wonderful!'

John ups the ante: 'Line up these skulls and tell a story to go with it. Publish that and then fight about it. That's what we do for a living.'

I sense scepticism on our side of the table. Is that really the way it goes, or is he just trying to get our goat?

'You folks can do it too, put together your own family tree,' John goes on. 'As long as you can spin a convincing yarn about the choices you make in chronology, and about the forks in the road.'

*   *   *

'THE GROUP STARES at the skulls in disbelief,' I read afterwards in Elfrieda's report. 'The topsy-turvy world lives on not only on the island of Flores, but also in the minds of the students.'

WE'RE STYMIED, SO WE ASK John de Vos for his view of things. Or perhaps we should say his story?

John places LB1 beside the skullcap of Dubois' ape-man and says that Flores Man is his direct descendant. That means he sees *Homo floresiensis* as a dwarfed *Homo erectus*. One million years ago, a delegation of robust Javanese hominids (of the type Dubois stumbled upon) must have crossed the Wallace Line. In the jungles of Flores, their descendants became more and more human and smaller, until – not so very long ago – they died out.

It was a scenario that would jibe perfectly with Father Verhoeven's discoveries.

'Did you know him?' I ask.

John nods. Of course: Father Verhoeven was a brilliant archaeologist, he had worked with him. As an atheist ('Don't go insulting me by asking whether I believe in God'), John had taken the former missionary along with him in 1988 to a conference on Sardinia with the Robinson Crusoeish title 'Early Man in Island Environments'. For more than a week, the two men kept each other company like an odd couple. Paula Hamerlinck went along too, and 'of course' John knew they were married. But wasn't that sort of a secret?

'Oh, no. They sent me a lovely marriage announcement at the time. They went on honeymoon to Flores.'

To our surprise, we hear that in 1990, the year Verhoeven died, John de Vos had gone on digging in the pits where the priest had stopped. He had literally picked up (the spade) where the other left off. Across from us, in other words, is a second 'MAN WHO DIDN'T DIG DEEP ENOUGH'.

Tragedy is a foxtail that sticks to your sleeve as soon as you pull it off your trousers. John tells us about his year-long efforts to get Father Verhoeven's pioneering work accepted in professional circles. He found an ally in his own doctoral supervisor, Professor Paul Sondaar of Utrecht, a bon vivant with a fear of flying, and also an honorary conservator at the Muséum d'Histoire Naturelle in Paris. Together they dug up a few stone side scrapers and hand axes from the substrate, which was also rife with stegodons. From this they drew the same conclusion Verhoeven had twenty years earlier: not only had elephants crossed the Wallace Line, but *Homo erectus* apparently had too. Who else, so long before the appearance of *Homo sapiens*, could have chipped away pieces of basalt to form tools?

Paul Sondaar and John de Vos expanded upon the discoveries made by Theo Verhoeven, who in 1970 had developed his earlier musings about primordial humans on Flores into a well-constructed theory. The islands east of the Wallace Line, he said, had already been colonised 750,000 years ago. By whom? By Dubois' Javanese *Homo erectus*. In the ecclesiastical journal *Anthropos*, Verhoeven had written most specifically about an *Uhrmensch*, a primal human, and its migration from Java to this place, *Einwanderung von Java her*, without anyone on earth believing his claims or paying them any heed at all.

Twenty years later, Sondaar and De Vos had trotted out the same scenario, but now with two additions of their own:

- Elephants have no need of a land bridge, as they can swim ('they possess a snorkel').

- *Homo erectus* was able to build rafts and rowed across the Wallace Line; it must have been more intelligent than had been supposed until then.

'We sent our article to *Science*, but it was sent back to us almost by return mail,' John says. 'They weren't going to touch it.'

The very next year, Paul Sondaar was fired. Budgetary cutbacks, it was claimed, and he hadn't published anything for a while. Sondaar fought back on two fronts: in the Netherlands by way of his lawyers, and on Flores with a battalion of villagers, whom he paid from his own pocket to go on digging at Liang Bua. He himself withdrew to write a children's book about a little Ice Age boy who lived on the mossy bottom of what is now the North Sea. The boy makes friends with an orphaned mammoth calf: they sleep close together to stay warm, then move off southwards to France, where the cavemen dwell.

The first big breakthrough came in 1998. 'Primitive humans conquered sea,' the international wire services reported. 'Early humans much smarter than we expected.' The only thing was, full credit was now given to a New Zealander working in Australia, Mike Morwood. The gist of Morwood's publication (in *Science*) deviated in no way from the two earlier studies, but it did make waves.

In the English-speaking press, Morwood made an additional speculation: Asiatic *Homo erectus* may have been capable of speech some 800,000 years ago. Sea travel by raft, after all, calls for not only coordination but also communication – and who knows, maybe this was accompanied by symbols and language?

Sondaar wrote furious letters, calling Morwood a 'paleo-fantast', but it got him nowhere. The sting lay in the fact that Mike Morwood had been given permission to work on Flores at the two Dutchmen's recommendation. John had personally applied for the grant to finance Morwood's follow-up study. 'With that, Mike went straight back to our sites.'

'Our sites?'

'Okay, Father Verhoeven's sites.'

'And he dug even deeper?'

'Mike went real deep. It was just plain dangerous. At Liang Bua: six metres, definitely. The walls of a well like that have to be shored up. Which he did, but still. If someone's working down at the bottom and a wall collapses, they'll be buried alive. We never wanted to take that risk.'

In the spring of 2003, Paul Sondaar died as the third 'MAN WHO DIDN'T DIG DEEP ENOUGH'. Of a brain tumour, embittered and spiteful beyond measure; he alienated himself even from his own circle of friends. Six months after Sondaar's funeral, in the autumn of 2003, Mike Morwood, now leading his own team, hoisted Flores Man to the surface, bone by bone. He had LB1 taken to room 109 of Hotel Sindha in Ruteng, the capital of Flores, where her remains were spread out on the bed like crown jewels on a velvet cushion. Morwood and his helpers referred to room 109 as 'the bone room'.

'I ran into him on Java that same year,' John says. 'Mike acted very secretively towards me. They had found something, but he didn't want to say what it was. I would hear about it as soon as it was published.'

That took a year, until late October 2004. They made the cover of *Nature*, and with that the entire international press. Every newspaper in every country.

'What am I supposed to say? Fantastic that Mike found this skull. Terrible that I didn't find it.' When pressed a bit, John de Vos refers to Mike Morwood as 'that bloody Brit'.

But he was from Australia, wasn't he?

'Yeah, or New Zealand – what difference does it make!'

THE GLORY LEFT to the Human Origins Group in Leiden: José's scoop in *Nature* about the zigzag line on the shell from the Dubois Collection.

Slides are shown, first of a misty landscape with rice fields, then of the gravel beaches along the Solo River, and finally close-ups of the shells of freshwater mussels, big as a man's hand. In some of them there is a little hole, 'drilled' by Dubois' ape-man in order to pry apart the two halves. The contents were eaten, while the shells themselves were filed down to make scrapers and knives. Out of boredom – or with a spark of creativity – one of the members of that *Homo erectus* clan must have picked up something sharp and scratched a drawing onto one of the shells.

To put it differently: the mussel shell was *signed*. And if humans are creatures who differentiate themselves from other animals by signifying their surroundings, then – voilà – this was a human. The first one, long before Adam.

José thinks that's nonsense. Too flattering for humankind. To the relief of art historian Els, José refuses to call the drawing 'art'. As far as she's concerned, this doesn't involve anything as weighty as 'the process of becoming human'. She shrugs off the suggestion that this etching might mark the transition from animal to human. 'I'll leave the interpretation up to others.'

For the first time, we turn our attention to the chunks of chocolate and the cookies – they've been lying there untouched all this time. But José isn't finished with us yet: she feels that we hold *Homo sapiens* in much too high regard. Consciously or not, we place our own species on a pedestal, even though there is no reason to do so. Bowerbirds create art too (bowers of sticks for their mating rituals), bees exhibit the division of labour, ants build bridges by seizing each other and holding on like rugby players in a scrum, dogs can hear higher frequencies than their masters: we humans are not nearly as wonderful as our self-conceit leads us to believe. Then she says: 'To me, a fish is every bit as competent as a human.'

This, instinctively, raises our hackles. People eat fish, and even though the occasional shark comes along and eats a person, there is no question of equality.

Bob gets into the discussion: 'Has a fish ever stopped to wonder about its origins?'

Touché, I think.

But objections like that make no impression on José. Suddenly she speaks on John's behalf as well: 'We're radical biologists who don't draw a single distinction between human and animal.'

I object: 'People are aware of their mortality, animals aren't.'

'What are you worried about?' says John. 'Another 20,000 years and we'll be extinct too.'

'People bury their dead,' I try again. 'And they want to know where they come from.'

While I view the staring skulls as evidence of this unique self-analysis (thanks, Bob: what other species tries to arrange its ancestors' bones in chronological order?), it occurs to me that I cannot simply get away with writing about deviation and norm. There's another question I have to deal with first: what grounds do we have for viewing ourselves as the measure of all things?

5

ONCE I THOUGHT THE WORLD of Leiden. The Crown Prince, now the King of the Netherlands, attended the university there. From the rather down-to-earth agricultural university at Wageningen, where I majored in 'tropical land development', I made the leap to Utrecht, Nijmegen, Amsterdam, to sit in on courses that touched on cultural anthropology, which weren't taught at our school on the banks of the Rhine. And so I travelled to Leiden as well. I went there for 'ethno-cinematography'. There was no question of simply registering and showing up; the limited number of spots were reserved for the most motivated students. I didn't think it was a hopeless ambition: my provincial background could be seen as lost ground, but I was ready and willing to make up for that.

The balloting session was held in a little theatre with a movie screen. We, the candidate ethno-cinematographers, took our seats. On the dais a little to one side was a table with a carafe of water and a glass, in the middle there was only a director's chair. For the same money, it may also have been a barstool or a kitchen chair. In any case, a slovenly-looking man in cowboy boots walked onto the stage,

silencing us at last by saying nothing and never looking at us. When
he finally sat down and ran his hand through his locks of hair, we
found ourselves eye-to-eye with a figure who looked for all the world
like Rainer Werner Fassbinder. Instead of welcoming us, he began to
insult us. We had, all of us sitting here in this theatre, a distorted view
of ethno-cinematography. There was nothing romantic about it – the
field demanded stamina.

Rainer Werner – I've forgotten the man's real name – had a glass
of water poured for him by his svelte assistant. She had the pallor of
a punk girl and wore black lipstick. Dedication, the man said: it all
came down to that. Ethno-cinematography was not filmmaking, it
was not about entertainment or beauty, nor was it about emancipa-
tion. It recorded. It registered. What did it register? Actions. Habits.
The daily grind in all its monotony. Whose actions, habits and grinds?
Those of the last primitive societies.

The ethnographic film, as it turned out, was related to the nature
documentary. Only now the film crew was not following a pride of
lions on the savanna, but *Homo sapiens* squatting together in the shade
of an acacia, or traipsing along behind broad-horned cattle.

The teacher snapped his fingers and the lights went out, he himself moved offstage, chair and all. What we were watching was the
making of an ethnographic film. First a landscape in black and white
with a few thatched clay huts – somewhere in Upper Volta. The professor, his female assistant and two villagers hand each other tools.
They are needed to build a blind: an observation platform in a rick-
ety kapok tree on a dusty village square; we hear the hammering, we
see a ladder affixed to the trunk. Our professor climbs up nimbly. He
hoists a movie camera and a few reels of film up behind him. In the
next shot – the way I remember it – we see him, shirt unbuttoned,

spying on the locals from his raised platform. Women in batik robes balancing jerrycans on their head, a boy on a bicycle.

'Who is the monkey in this picture?'

I address the question to my students, rhetorically. (And no, I did not study ethno-cinematography at Leiden.) What I want to know now is whether there are any ethnographic films about Flores in the early years of the twentieth century. In order to write, we need images, I say. What did the island look like when the missionaries from Steyl first saw it, upon their arrival in 1916? On the whiteboard in our old, familiar classroom in the Humanities building, I note the names of Els and Marijn: they volunteered to go in search of old footage of Flores.

Soft-spoken Elizabeth signs on for the sub-study on pygmies.

'And who's going to get us the nitty-gritty on Lilliputians, starting with the island in *Gulliver's Travels*?'

The rest of the students hold off; they are waiting to hear whether I have other trails for them to follow. But then upcoming classics scholar Thom raises his hand; he offers to find out, along with Elizabeth, what the term 'dwarf people' was supposed to designate, and how this relates to Father Verhoeven's proto-Negrito from the cave at Liang Toge.

To help Thom and Elizabeth, I give them a copy of *The Negritos of the Eastern Little Sunda Islands*, by W. Keers. As leader of an expedition in 1937 and 1938, Keers had measured thousands of natives of Flores and the neighbouring islands. The Negritos (literally: 'little black men', the diminutive form of the Spanish *negro*) were the prime focus of interest among the physical anthropologists of the day. Were these slender Asians with their short limbs actually related to the African pygmies or not? In 1957, the Associated Press wire service quoted Father Verhoeven as saying: 'The Negrito human is related to the ancient Mongoloid human, who was one of the precursors of the yellow race.'

Concerning Negrito expert W. Keers, I knew only that her first name was Wilhelmina and that she was not married (in a treatise concerning Sumatran fireflies, she was spoken of as 'Miss'). Not only had she applied her calipers to nostrils and earlobes, but she often took blood samples as well. 'Ethno-sanguistics', such research was called.

A gust of indignation rises from the class. 'Vampirism' seems to my students to be a more fitting name for it.

What I admire about Wilhelmina Keers, however, besides her having been a woman in a man's world, is that she had looked beyond stature alone. Yes, pygmies are small (shorter than 150 centimetres is the general standard) and Negritos are barely a hand's-breadth taller (150 to 160 centimetres). But that's plain as day, isn't it? That's no reason to carry out a study. 'Is it merely their shortness that sets them apart?' the physical anthropologist asked herself. In addition to blood samples, Ms Keers also collected fingerprints and conducted taste tests: did pygmies/Negritos possess receptor cells for bitterness?

Philosophy student Lian cuts short my prattling. 'If you ask me, what it's about is this: when do we view small creatures as aliens?' With one question she has linked the notion of deviance and norm to that of otherness, and, by extension, to exclusion.

Lian is active in Leiden's student cabaret group: she has just given her first solo performance on stage. Now she suggests that we examine stories about giants and dwarfs, in search of the implicit moral behind them.

'Tom Thumb,' someone shouts.

'Snow White and the Seven Dwarfs.'

And also: 'David and Goliath.'

For the purposes of our quest, there can be no harm in examining the connotations normally attached to big and small. What exactly are they: value judgements along the lines of higher/lower? Superior/inferior? Anyone attempting to describe who 'we humans' are is in need of a 'them'. The ancient Greeks used the – immortal – gods to that end, but it's much more practical to compare oneself to the animals. Roughly 100 per cent of the world's population see humans as superior to beasts. They may have the Lion King, but we have the lion tamer. To some we are stewards appointed by God; to others we are simply at the top of the food chain.

Even those who recognise that we are animals as well continue to see themselves and their congeners as the superlative stage of the animal.

To get a better view of the dividing line between human and animal, I add 'the missing link' to the list of assignments for further study. In the late nineteenth century, this was an idea that echoed throughout Europe. *Le lien manquant. Die fehlende glied.* Newspapers and magazines speculated wildly about this unknown transitional form between apes and *Homo sapiens*. The appearance of this half-animal, half-human could, once found and displayed, cast light on the transition from animal to human. I told the class that Ernst Haeckel, a German zoologist and dead ringer for Karl Marx, imagined the missing link in 1868 as a 'speechless prehuman'. The single great criterion that made humans human was, in Haeckel's view, the ability to speak.

We agree that our search for the missing link will be a group effort, as will the drafting of a list of 'most commonly mentioned distinctions between human and animal'. In addition to speech and the ability to use fire, the capacity for abstract thought, artistic ability, self-awareness and cognisance of mortality have to be included as well.

My final and apparently most straightforward research chore goes
to seventeen-year-old Manola. She had skipped half of secondary school
and is now doing a double major at Leiden, in French and Dutch. Manola
chooses to investigate the mysterious report of a village in Central Flores,
said to be home to seventy-seven dwarf families. 'Rampasasa' is its name,
but it appears neither in the *Times* atlas nor on Google Maps. On the
internet there is a reference to an article in the *Jakarta Post* from 2005,
in which Professor Teuku Jacob announced that four out of every five
adults in Rampasasa were no taller than 150 centimetres. Jacob described
them as 'dwarfish'. According to him, these rather diminutive people are
the direct descendants of LB1, who, at 104 centimetres, was a completely
normal *Homo sapiens*. It sounded as though he were saying: 'Rampasasa
is inhabited by hobbits, but there is nothing so unusual about that.'

Like Lian, Manola is a small girl. She has grown up conversant
in Bahasa Indonesia, a standardised form of Malay. Her father was
born in former Dutch New Guinea. As a Moluccan Reformed pastor,
he delivers his sermons in Indonesian every Sunday and so can help
where necessary with the translation of Indonesian texts.

Despite the tentativeness of our efforts at this stage, a team spirit
is growing. The horseshoe of desks is losing some of its rigidity. What
I had been hoping for has happened: our clubhouse in the block-like
Van Eyck building is turning into a workshop.

Besides the content, the point now is to find a narrative form. In
writer's jargon: the time has come to switch from 'the what' to 'the
how'. To that end, I fall back on the motif of 'the making of'. In every
account of a search, I lecture, one can include the course of events
itself in one's reportage by way of a thin, red line, in that setbacks are
more valuable than victories. They're like enriched uranium – they
have a higher dramatic core weight.

'What doesn't kill you makes you stronger,' is how Mariëlle sums up my point.

'Something like that, yes.' Then I talk about the phenomenon of wrong tracks. In a true-to-life story there is no reason not to follow those for a bit: wandering or backing the wrong horse is part of reality: it can provide a quest with character and depth.

For the same reason, the person of the 'unreliable narrator', known to us from fiction, also deserves a place in the reportage. In the wild, after all, there are probably more unreliable than reliable narrators walking around.

I hesitate about bringing the frame story into it too. I could easily do so, by making use of 'Sjaalman's package'. Perhaps it is my job to point that out as well: *Max Havelaar* is seen as the gold standard of modern Dutch literature. Multatuli created a story within a story: the protracted grumblings of Droogstoppel, dealer in coffee, as the frame within which his German apprentice, Stern, transcribed the documents that had come to them in Sjaalman's package. Finally, the author himself comes crashing into the story: 'Yes, I, Multatuli, "he who has borne a great deal", am the one addressing you now.'

But I pass on that. I've done enough hinting at frame stories: I keep notes on the things we talk about, and after each session I photograph the scribbles and arrows on the whiteboard. Beside each name on the class roster, I write little profiles ('gruff', 'dots the i's with little circles', 'hyper') as one would for potential characters in a story. I want to hold on to the possibility of giving my students active roles in my book. Who knows, maybe I might come up with a structure in which a 'making of' coincides with a frame narrative?

*   *   *

BACK HOME, AT MY LAPTOP, surfing my way to the mythical island of Lilliput, I stumble upon the actual 'Kingdom of Little People'. This turns out to be an amusement park in the Chinese province of Yunnan, also known as the 'Dwarf Empire'. I read that scores of people with dwarfism live and work there – twice a day they put on a performance for paying visitors. In 2014, a Flemish photographer had documented life there behind the scenes. In one photo, a pair of park guards look like uniformed giants as one of them holds up a tiny woman in a bridal gown. She must weigh almost nothing, because she's standing on his outstretched palm; the security man displays her like a doll he's just won in a shooting gallery.

The little bride earns a living by looking cute. Being little makes big people think of children. Of innocence, care and play. Just like with young animals, they feel instant tenderness. Giants, on the other hand, are stupid and clumsy, and have feet of clay.

From our first, summary tour of sagas and myths, we note that behemoths are more often brutes than softies. David slays Goliath, the little ones win our support and favour – a case of sympathy for the underdog. This does nothing to detract, however, from the fact that the Chinese Dwarf Empire reminds one, and powerfully so, of the old European freakshows – like the 'Märchenstadt Lilliput' in Berlin, which failed to survive Hitler's ascendancy. No matter how much entertainment they provided, Nazi ideology saw them as an aberration that threatened the purity of the Aryan race.

The simple qualification 'sick' was enough to set in motion the machinery of banishment. Lepers were sent to leper colonies, albinos in Africa were tried as witches. Application of the label 'pathological case' was an oft-used prelude to condemnation. Teuku Jacob applied this tactic to LB1 when he stated that Flo suffered from the same deformity (microcephaly) as babies whose mothers had been infected with the Zika virus. Wasn't that a much more likely explanation? Jacob received vocal support from half a dozen physicians and fellow anthropologists from Johannesburg to Adelaide, united under the name 'the Pathology Group'. Some felt that Flo had remained small due to an iodine deficiency (cretinism); others said it was due to an insensitivity to growth hormones (Laron syndrome). All of the Pathology Group's efforts were directed at disqualifying her as a physically sound member of a separate human species.

Which brings us to an essential difference between dwarfs and pygmies. With the former, one can speak of a growth disorder, a defect; with pygmies, however, limited growth is a shared trait: it is the norm.

At our next session, Thom gives a presentation dealing with a 'blood hunt' in the Belgian Congo, organised in 1933: a semi-scientific contest to see who could collect the most samples of pygmy blood during a three-month period. At the time, Thom explained, there was a notion that the blood of pygmies had a different chemical composition from that of the rest of humankind. A chemist from Utrecht, travelling about by sedan chair, was the winner, despite the fact that many of his 'little brown friends' had fled from his hypodermic needle.

Physical anthropologists, having caught 'pygmy fever', set out on a series of forays into the African and Asian interiors. In Papua New Guinea, a Dutch expedition discovered a tribe at the foot of Mount Goliath with an average height of 149 centimetres. In accordance with a truly colonial brand of humour, they were christened the 'Goliath Pygmies'.

To this Wilhelmina Keers added the discovery of a group of 'little Negritos' in Central Flores. Not far from the Inerie volcano she had found this group, recognisable as such by their black, curly hair and colour 4 on her *Augenfarbentafel*, or eye colour table. They were not quite short enough to qualify as pygmies. Ms Keers believed they were descendants of the very first humans to have set foot on Flores.

I thank Thom for his presentation, and am just about to say that the physical anthropologists seem to have been on the wrong track, when Elizabeth interrupts me by murmuring something inaudible. Everyone looks at her.

'I worked on this presentation too,' she repeats, this time with a little more volume.

I'm about to make good by thanking her as well, but the professor of modern Dutch literature beats me to it. 'Better get used to it, Elizabeth,' she comments. 'That's the way things go in the real world.'

WHY IS IT that I prefer to write reportages rather than novels? My simplest answer: because reality always wins hands down over my own fantasy. Time and time again I stumble upon true stories that make such an incredible impression that were they to receive a liberal coating of fiction, they would immediately lose their last little sliver of credibility.

When asked for an example, I mention the plan launched by a Dutch anthropologist in the Congo to cross female gorillas and chimpanzees with 'Negroes', in an attempt to resuscitate the missing link. It was never carried out, yet Queen Wilhelmina of the Netherlands actually provided funding for the experiment in 1907.

The raw material provided by reality seems so baroque to me that I feel no need to add more curlicues to it. Take, for example, the fact that children baptised on Flores were given a 'Catholic mohawk': the Steyl missionaries made a sport of shaving the heads of their youngest disciples, except for one spot in the shape of a cross on the top of the skull. What that Roman mohawk looked like and the easiest way to make it is something we learned from old black-and-white footage.

It was, as it turns out, not the ethnographers but the missionaries who pioneered ethno-cinematography on Flores. Flecked and jittery but recently restored and digitised, Els' and Marijn's find consists of several 'mission films' from the Society of the Divine Word.

When we project a film from YouTube, Classroom 0.04 transforms into a movie theatre. We see waving palm trees, pile dwellings with raffia roofs and brown-skinned altar boys in 1920, trotting along with a lampshade-like parasol behind a lily-white priest. As a teenager, Theodor Verhoeven must have seen this *Flores-film, a voyage to Insulinde and the mission on Flores*. One of his teachers at the seminary, in an era later referred to as 'The Great Age of Missions', had formerly driven around the Catholic south of the Netherlands in a Model T Ford with FLORES FILM emblazoned on the side.

'A work of art of great ethnological value,' wrote the local *De Maasbode* daily paper. 'Deeply touching in portent.'

'Highly ambitious. They have succeeded in putting together a motion picture that provides a perfect view of country and people,' ruled the *Koloniaal Weekblad*.

Around 1910, Pope Pius X had threatened his clerics with excommunication were they to visit 'movie houses'. But even if it did give young men and women an opportunity to meet up in darkness, the medium still cast a spell more powerful than the Holy Spirit.

In 1929, a priest-director and a priest-cameraman from Limburg province travelled to New York to attend the Institute for Photography and Cinematography. After receiving their diplomas, they travelled across the continent, visited the Hollywood studios and signed onto ships, taking them to Tahiti and then on to the Philippines and Java, before they returned to their mission on Flores to make *Ria Rago: The Heroine of the Ndona Valley.*

It is a feature film. One with actors, a script and a plot: a Catholic girl, Ria Rago, is given in marriage to a heathen; she rebels and seeks shelter at the mission station. We see her, weepy and crestfallen, standing before a tall man dressed all in white. An intertitle appears: 'Toewan (Padre), my father wants to marry me off to Dapo.'

The missionary leans towards her and strokes his goatee. 'To Dapo, who aspires to being a Muslim and already has a wife?'

The priest sends a curate to talk her parents out of it, but to no avail. Ria Rago's family kidnaps her and gives her a beating – just to teach her a lesson. Bleeding like a wounded animal, she escapes again to the mission hospital, where she dies holding a crucifix, but not before forgiving her parents with her dying breath.

At its first showing, still on Flores, there was an audience of some 2500 people. One of the priests wrote the program to go with the film:

> *Rugged and hostile is the landscape of Lio, backbone of the island of Flores, like a stack of mountains; the country lies squeezed on high between two seas, the mountains hanging in the midst of the countryside, shoulder to shoulder and back to back, like giants with their feet braced deep in the sea, pushing one against the other.*

Off the coast, 'like a boat covered in palms adrift', lies the island of Ende, 'green paradise in the infinity of blue'.

> *From that island, Islam came to the coast and made its way to Lio, where it continues to push further and further into the mountains.*

Amid great globs of proselytical fever, calculation and meticulousness lie hidden here, in both film and program. The claims made concerning 'native custom' are based on careful study of the *adat*, or common law. The buffalo hunt seen in *Ria Rago* (Dapo is required to pay one buffalo by way of a dowry) is pure ethno-cinematography. This is exactly how the men of Flores, armed with bamboo staves, would cut a buffalo and her calf away from the herd, surround and capture her.

The priests carry out anthropological fieldwork. Their order explicitly promotes the gathering of knowledge. Among the missionaries are linguists, ornithologists and, later, archaeologists. 'For with much wisdom comes much sorrow,' Ecclesiastes says. But the motto here is: 'Knowledge is power.' The better the priests understand the *adat*, the better the light in which they can present their own set of morals.

Yonder on Flores, the missionaries toss up a dam against Islam; at home, beside the River Meuse, the home front does its best to halt the godless ideas of Darwin. The conventuals at Steyl add their own facet of meaning to the term 'evolution': Christians are *evolved* heathens; they are one rung higher up on the ladder. Those not yet converted are children, they still have to grow – *that* is the evolutionary path followed by humankind.

Nine out of ten inhabitants of Flores turn out to be baptised. The score of converts is remarkably high. Within the biggest Muslim country in the world, Indonesia, the population of Flores rises in the 1960s to more than a million, of whom more than 80 per cent are registered as members of a single religion: Roman Catholicism. Apparently, due to the assiduity of the Steyl priests, today one finds almost no one in the island's interior who will not eat pork.

But still we have our doubts about whether this insular Catholicism went any deeper than outward signs, like the Roman mohawk. The Bible may well have made its way into Flores' remotest valleys, but the indigenous world of the gods seems to have been anything but overrun by it: it had at best only moved up a little to make room. Young Manola, the product of a staunch Moluccan Protestant clan, reports to us about a widespread heathen belief on Flores. No village by the name of 'Rampasasa' can be found, but in its place Manola has discovered a wealth of narratives about 'little hairy forest demons'. *Ebu Gogo*, the creatures are called, and almost everyone on Flores believes they exist. If you happen to see one, it is usually because it has snuck in to steal an ear of corn from the fire. They look like humans, but they are not.

In our group's Dropbox, Manola posts a report by Rokus Awe Due, an archaeologist and one of Father Verhoeven's former students. From the time he was a child, Rokus always made a point of being there whenever his mentor started a dig, 'even if I had to walk a week to get there'. Years later, in 2003, he was witness to the emergence of LB1 from the cold cave at Liang Bua.

Rokus remembered how his father used to warn him about the *Ebu Gogo*. 'When it rained,' Manola tells us, 'he wasn't allowed to go outside; otherwise, the *Ebu Gogo* would get him.' These mysterious creatures are 'a sort of animal that can run very fast'. Rokus summed up their traits as follows:

- hairy
- barely one metre in height (!)
- evil.

Stories like this were first written down by – how could it be otherwise – one of the Steyl priests. This was Jilis Verheijen (1908–1997) from Ooij, a parish village on the Rhine.

Verheijen, who combined his priesthood with lexicography and ornithology, was a regular companion of Theodor Verhoeven. His passion for word-collecting also extended to birdsong. This resulted in long lists of bird names: red-necked phalarope, flame-breasted sunbird, black-winged kite, coconut lorikeet. Pure poetry, but then written in Manggarai or Ngadha. On the basis of a large, speckled egg given to him by Father Mommersteeg, Verheijen claimed to have discovered a giant version of the cuckoo, *Scythrops novaehollandiae*, which few ornithologists suspected might be found on Flores.

Verheijen's legacy covered 2056 pages. One section of these Manggarai texts consists of an anthology of fables and legends. Manola promises that she will try – with her father's help – to skim these collected folktales for signs of the little creatures that resembled Flores Man.

Where did these creatures come from? Could they have been simply figments of the imagination? Or were these accounts based on something like real documentation, and were these the actual descendants of LB1: small clusters of *Homo floresiensis* who had seemingly survived the volcanic eruption 12,000 years ago, in any case until the arrival and rise of *Homo sapiens*?

It is not entirely unthinkable that oral legends concerning little, hairy forest demons might hark back to the last hominids to exist side-by-side with modern humans.

Perhaps we have dismissed Wilhelmina Keers a bit too flippantly. In a footnote overlooked at first, it turns out that she had used folktales to back up her assumptions about the origins of the 'little Negritos'.

'These are important stories, in view of the fact that they talk about a population of hairy indigenous people who lost that trait [hairiness] when they became more civilised.'

In a kampong by the name of Tjibal, for example, Wilhelmina Keers noted that a tuft of grey human hair had been preserved for who knows how long. 'Flores used to be covered with jungle, which was crawling with wild animals,' the oldest of the village elders told her. 'There were no people.' Suddenly, though, a man and two women appeared. The man was covered in hair from head to toe, except for his face. After the three of them had lived for about ten years in great hardship, a strange woman appeared from the coast. She showed them how to build a fire and then disappeared. As soon as the man tried it himself, his grey hair was scorched and fell out.

The *Ebu Gogo* are also included on Manola's list as primitive hominids who have no knowledge of fire. They eat their food raw. Both males and females are covered in hair and 'stink like billy goats'. They are terrified of dogs.

Another point that rarely goes unmentioned: the women have extremely long breasts, which they can toss crisscross over their shoulders whenever that suits them.

Until only a few generations ago, one story has it, the *Ebu Gogo* lived in a cave not far from the village of Ola Bula. The villagers suffered so severely under the pillaging of their crops that they decided to eradicate the little people. To do so, they cut five hundred palm fronds and laid them by the cave's entrance, 'for them to sleep on'. The ostensible peace offering was accepted, but the *Ebu Gogo* actually used the fronds as clothing. In any case, the plan hatched by the people of Ola Bula succeeded; they set fire to the cave, and the flames spread rapidly because of the dried fronds. All the *Ebu Gogo* were killed in the conflagration, except for a man and a woman, who escaped and headed in the direction of the Inerie volcano.

'Ever since then, maggots come crawling out of the cave, then die in the sunlight. Dried-out worms have even been found as far as a kilometre from the entrance.'

On the whiteboard, I write three questions prompted directly by this legend:

1. Did we, *Homo sapiens*, wipe out Flores Man?
2. Were Wilhelmina Keers' 'little Negritos' (as well as Teuku Jacob's dwarf families at Rampasasa, perhaps) descendants of a cross between *Homo sapiens* and LB1?
3. Are there, perhaps, still a few *Homo floresiensis* wandering around, somewhere on the wooded, volcanic slopes of Flores?

If the answer to that last question is yes, then there exists a living alternative by which to take the measure of 'humanness' – and we can no longer see ourselves as the only benchmark.

6

Eijsden, 18 December 2016. The Dubois Society is out for a stroll along the high embankment above the Meuse. The weather is mushy, and the air is cold and clammy. I've signed up as a member – I am one of them. Even though it's not raining, the cobblestones glisten with moisture. We go down a set of steps in the embankment wall to a nineteenth-century boundary post, which marks the division of the Low Countries in 1830. Right beside it is an outdoor cafe with olive trees in pots; pending any further rise in the earth's temperature, the Mediterranean foliage has been wrapped in canvas covers.

Down there in the mud is the start of the Eijsder Beemden nature study route, which the hiker has no choice but to share with long-haired cattle. On the far shore: Belgium.

'Ternaaien,' the society's treasurer says.

I still have a great deal to learn.

'The village there, on the other side.' In summer, he tells me, you can take a ferry, reserved for cyclists and pedestrians, to French-speaking Ternaaien/Lanaye, but today the pontoon docks are empty.

Two kilometres downstream one can see the silhouette of Mount Saint Peter. 'As a boy, that was Eugène's view from his bedroom window.'

As the novice among the Dubois disciples, I'm the one asking questions in amazement. The Dutch and Belgian flags on that finger of land, there, past where the cows are – why are they flying side-by-side?

'Ah,' a voice says from above. 'The border realignment.' The treasurer's wife has stayed behind on the upper quay; her dress shoes aren't suited to the path. 'They held a little ceremony here,' she says, 'at the same time as the big ceremony in the palace in Amsterdam.'

I'll be damned, yeah, the border realignment. The news about that had unfolded as a fairytale: two royal couples bestowing land on each other, back and forth, without a single soldier having to die first. 'All in peace and harmony,' King Willem-Alexander said during the treaty ceremony. 'No need for a ten-day crusade.'

It was all about two marshy, uninhabited enclaves. The land tongues lay on either side of the Meuse. The first (the size of fourteen soccer pitches) was Belgian, but attached to the Dutch shore; the other (four soccer pitches) was Dutch but lay on the far side of the river.

Difficult as this may be to explain, the *petite histoire* laid out for my edification in Eijsden is even odder. It all began – and I am not making this up – with the discovery of a body. A blue, swollen torso with limbs, no head. An angler found the naked corpse at the far end of the very path we are following, on the finger of land, out past the flags. Police from Eijsden were not allowed to go there: Dutch constables had no jurisdiction on the territory of the sovereign state of Belgium. But the Ternaaien police, once informed of the situation, were not easily able to reach the crime scene either. Along with a forensic team, they had to cross the fast-moving river in a motor launch and, in the absence of any dock or bollard, jump ashore once they were within a few feet of land.

'And every summer the folks from the Moustache Club came here and took possession of it.'

The Moustache Club?

'Fellows in leather, with moustaches. From Antwerp.'

Once a year, the hirsute men declared the land tongue to be the lawless republic of 'Moustachio'. What they did there, however, was nothing compared with the rites of spring performed by the throngs of nude males who frequented the little peninsula.

'Its always been a gay cruising ground too,' the treasurer's wife explains.

The riverside beaches surrounded by thickets, where the pharmacist Jean Joseph Dubois and his sons Eugène and Victor had gone to pick medicinal herbs, are now littered with soggy condoms. Amid them, ancient cattle breeds mill about like demonstrators protesting for a better society. Not everyone is pleased with the situation. On TV Limburg, one schoolteacher from Eijsden reported that during a biology field trip she and her class had stumbled upon copulating men.

THUS FAR HAS evolution brought us. Or is it civilisation, in fact? It wasn't that long ago that Western psychiatry would have prescribed castration – chemical or otherwise – for moustachioed men like that. Or a lobotomy. These days, anyone using a word like 'deviance' to describe them can count on a sturdy public bashing.

But things tipped. Or turned themselves inside out. This brings me to my next point, that of the central question as moving target. The thing you are pursuing in a discursive reportage moves as you go along; it shifts. Sometimes that happens gradually, like a heavenly body in the eyepiece of a telescope; sometimes in fits and starts, and as fast as the little tin rabbit in a shooting gallery. In both cases, the point is to promptly move along with it.

Beating the pavement was a precondition, a requirement. The novel can get by without fieldwork, the reportage cannot.

Exploring Eijsden in the slipstream of the Dubois Society gave me a place to start. Who was interested these days in commemorating an evangelist of Darwinism who had rocked public opinion around the turn of the next-to-last century? Had Eugène Dubois permanently altered our view of humanity? And, if so, how?

Of the forty-one students who had signed up for my course in the autumn, sixteen were still with me by mid-December. It was precisely at this point that exam week began, and that had quashed my plans for a second group excursion. Mariëlle, who already had her master's degree and therefore had no exams, was the only one who'd set off with me for the southernmost frontier station in the Netherlands.

WE GATHER AT the churchgoing hour in the old Ursulines convent, a severe building of oxblood-red brick. It was this convent Eugène Dubois' sister Marie had entered as a nun in 1876, to his great chagrin. While he was off on the other side of the world, she was passing her days in the same street where she was born. Our reason for getting together in the former chapel of the Ursuline Sisters, however, is not Marie but her anti-religious brother.

The morning's program includes a tribute to John de Vos as a lifetime member of the Dubois Society. Wearing a checked shirt and leaning one elbow casually on his stool, he stands at the front, beneath the stained-glass windows. While above his head John the Baptist is bathed in a beam of heavenly light, John de Vos gives a talk about the first Dutch fossil hunters. To my surprise, he also brings up the Himalayan yeti, which the comic books call the 'Abominable Snowman'. I had always assumed it to be an imaginary creature, but it turns out I was wrong. The yeti is no shadowy projection of our innermost fears. *Gigantopithecus* – giant ape, quite literally – is its scientific name. It was, in fact, Professor Von Koenigswald who first identified this largest of all primates, on the basis of a couple of 'dragon's teeth' bought from a shaman in China in 1935. Decades later a few molars were discovered as well, and part of a jawbone. John describes *Gigantopithecus* as a three-metre-tall bamboo eater. Had he not become extinct long ago, he might have evolved into a 500-kilogram, ten-foot-tall hominid.

We, his listeners, ask ourselves: and then what? Would *Homo sapiens* have been any match for this *Homo giganticus*? (Imagine if there were seven billion of them, all in need of a place to live . . .)

John de Vos steps away from the lectern and, to loud applause, receives his certificate of appreciation from the Dubois Society. And that's not all. For his forty years as curator of the Dubois Collection, John is given a wine crate. When he turns it around to show the audience, we see that there is no bottle of wine in it. Instead, it holds a bronze statuette of the Javanese ape-man. Mariëlle and I look at each other. After the ceremony we go up and congratulate John and his wife, Rita, who takes the certificate and the statuette from her husband's hands.

'Are you going to hang that certificate on the wall above your bed?' Mariëlle asks.

'You must be kidding,' John replies.

Rita – a row of piercings in each earlobe – takes the ape-man out of his box. 'I'm afraid it's going to be up to me to dust this thing.'

The assembly moves to the refectory for coffee and a slice of flan. Our route takes us through tiled convent halls, where an exhibition on human evolution is being held. As soon as the last nun left the Ursuline convent, the niches with their Madonnas and crucifixes had been emptied. Now they are occupied by all varieties of hominid skulls.

'A total makeover,' Mariëlle remarks. We were both thinking of Els, whose absence spared her the sight of this iconoclastic verve.

The finest niche is dedicated to Eijsden's most famous son. There he is in the flesh, or at least in the form of a mannequin in a white shirt and grey collarless jacket. With his protruding belly, Eugéne Dubois looks the very picture of the merry innkeeper. A few steps further, past a stack of boxes with statuettes of the Javanese ape-man (on sale for 89 euros), the wall of the cloister is decorated with a life-size, handwritten Dubois family tree.

While we are examining it, a gentleman in a three-piece suit comes up behind us. He introduces himself as a Dubois. 'Victor Dubois, Eugène's brother, was my grandfather,' Dubois says, pointing to one of the lateral branches. 'But he was only a law-abiding family doctor in Venlo.'

Suddenly I am struck by the feeling of having missed something. What has made Dubois such a cult figure – at least among his disciples?

We direct the question at John de Vos, when we sit down beside him in the refectory.

'I thought I'd explained all that to you in Leiden? Or has it slipped your mind?' John sets his coffee aside to drill us on the importance of the skull found by Dubois. He begins with the term 'holotype', used to denote a fossil considered normative for a given species. During an outbreak of Ebola, for example, medical science goes looking for Patient Zero – the first victim, and the source of the epidemic. Analogously, anthro-paleologists tend to see the first found specimen of a certain hominid species as their benchmark.

Holotype. The word alone is enough to make my imagination soar. It has a magical ring to it, I say.

John comments drily that he himself has no real nose for magic. 'But I've also noticed among my colleagues a certain awe for the holotype.' In 1982, he tells us, during a conference in Nice, he was approached between lectures by a group of Americans from the Natural History Museum in New York. They were busy setting up a special show under the unassuming title *Ancestors*. The unicity of the special exhibition lay in this: for the first time in history, the museum's visitors would get to see the original skulls. Anyone wandering in from Central Park would find themselves eye-to-eye with the 'real stuff', not replicas. John, the Americans were aware, was the guardian of Dubois' skullcap – and there was no way that the holotype for *Homo erectus* could be left out.

'They invited me to New York for a briefing,' John says. 'At the hotel desk, I was given a suite.' In the other suites, as it turned out, were professional colleagues who, just like him, guarded a strongbox with a hominid skull – from Nairobi, Addis Ababa, Johannesburg.

Raymond Dart was there too: the man who had discovered (in 1924!) the more-than-two-million-year-old 'Taung Child' in a South African pit mine. 'Raymond Dart was in his nineties by then,' John said. 'They had to carry him.'

The finder of Lucy, the 'ape-woman' of Ethiopia, had also been invited. As with the Taung Child, Lucy's absence in New York would have been unthinkable.

John de Vos and his colleagues were smothered in promises. A limousine would be waiting for each of them at JFK Airport. As soon as the curator climbed in with his priceless hand luggage, they would leave under police escort for Central Park West and 79th Street. The couriers would be allowed to place the skulls themselves in a bombproof and bulletproof display case. The deal they made was solid, airtight.

But fourteen days before John was to return to New York for the exhibition, the loan was called off. Not by America, where creationists ('STOP THE BLASPHEMY') and anti-racism demonstrators ('APARTHEID SKULLS NOT WELCOME HERE') were already painting their protest signs, but by the Netherlands. Foreign affairs minister Max van der Stoel felt the American guarantees were insufficient.

'But what was the problem?' I ask.

'They were afraid that Indonesia would claim the fossils.' A rumour was making the rounds, John says, that Professor Teuku Jacob was planning to retrieve them on Jakarta's behalf.

AFTER THE COFFEE and pastry, we set out afoot through Eijsden. Cater-cornered to the Ursulines convent, we stop in front of a rambling building reminiscent of a farmhouse. Here are both the pharmacy and the house where Eugène, his two sisters and his brother were born. Peeking through a window is out of the question: they are all hung with bright-yellow posters reading 'STOP TIHANGE', with the black warning sign for radioactive risk, a protest directed against the three nuclear power plants upstream along the Belgian Meuse.

Despite the weight of John's New York anecdote, I continue to ask myself what it is that makes a special person out of someone who finds something unique. Newton was the first to understand the force of gravity. Tesla invented the AC motor. But what could Dubois take credit for?

According to John, I am still missing the point. Why don't I compare him to Columbus instead? 'Dubois was not only the first finder, but also the first seeker.' John characterises him as a man with a mission, a lone wolf who braved hell and high water to find material evidence that the human race was not created, but arose.

We continue our walk at quayside, looking out over the Meuse and the greenery. 'We're talking here about the nineteenth century, right?' John goes on. 'You couldn't find a cleric who didn't recoil in horror at the word "evolution".'

I am starting to view Dubois differently. If I disregard the folklore of this morning's program, my mind's eye actually does see a courageous figure who boards a ship for the Dutch Indies as an explorer, but an explorer of prehistory. The territory across which he meant to blaze a trail was the past. Like the geologist, he possessed the gift of being able to see beneath the ploughed fields and rice paddies entire 'continents', with names like Pliocene and Pleistocene. He had come to Java to search through layers of time for traces of human and pre-human life. What he was asking himself as he did so ('Where do we come from?') was nothing new. But unlike almost everyone else before him, he didn't look to heaven when he asked that question, but at the earth.

Mariëlle, who had a Protestant upbringing just like mine, points out the irony in the fact that Dubois had made good at least in part on the biblical directive 'seek and ye shall find'.

John goes no further than to observe: 'He pulled it off, that's all.'

*   *   *

BEFORE STARTING IN on our shrimp rissoles and white bread at bistro La Meuse, we pose for a group portrait. The hale and hearty nature of our Limburg hosts and hostesses is evident not only from their attire – neatly informal – but also from their relaxed bearing. All the more unexpected, therefore, is the rigour with which the board members of the Dubois Society defend their godless, biological-reductionist worldview.

At the table, our talk centres on whether traits such as courtesy, sadism, magnanimity, unholy glee and generosity are actually seen in the animal kingdom – that is, if one doesn't count *Homo sapiens* as a member of that kingdom. I pose the question to the Dutch translator of Darwin's work, a Fleming with Leonard Bernstein–style hair, a lock of which keeps falling over his forehead whenever he shakes his head in indignation.

'Translating *On the Origin of Species* was the highpoint of my life,' he says with conviction, as though he expects no further summits. He had stuck to the first edition, from November 1859, he says. During Darwin's lifetime, five more editions had followed, to which the author had added a closing paragraph dealing with 'the Creator'. The Darwin translator knows exactly how that sentence went, the one he had *not* translated:

> There is grandeur in this view of life, with its several powers, having been originally breathed into a few forms or into one; and that, whilst this planet has gone cycling on according to the fixed law of gravity, from so simple a beginning endless forms most beautiful and most wonderful have been, and are being, evolved.

It sounds religious.

'Exactly! Darwin was pressured into writing that. Letters he wrote later on show that he regretted it for the rest of his life.'

While he looks for a business card to hand me, I ask him the question that is bothering me: 'People are animals, fine. But don't they exhibit – don't *we* exhibit – traits Darwinism can't explain?'

'Which traits would those be?' He looks me in the eye; the lock of hair has slipped beneath the rim of his spectacles.

'People kill themselves.' The words barely cross my lips when in my mind's eye I see lemmings running towards a cliff to throw themselves into the sea. I am about to correct myself by saying: 'There are members of our species who blow themselves up for a cause.' But the Darwin translator's neighbour interrupts me. He leans towards me and starts talking about lemmings.

A few more members of the Dubois Society – the men had gathered around one end of the table, the women around the other – join in the conversation. Beakers of monastery beer are set before us; a bottle of Sancerre in a cooler is for the ladies. The wallpaper in the dining room bears a gilded text:

> The Meuse hides / our differences unique
> The Meuse binds / the languages we speak

The cultured impression we may be making on the waiters at La Meuse does nothing to change the fact that everyone at the table, with the exception of Mariëlle and me, lives in the conviction that there is absolutely no difference between us and the apes. *Homo sapiens* is simply a species of primate within the family of hominids, a tiny niche within the animal kingdom that we share with the chimpanzee, the bonobo, the gorilla and the orangutan. Ape number five is who we are, with no claim at all to a status apart.

'People aren't aliens,' John de Vos chimes in. 'We all come from the same planet.'

Someone begins citing Dubois: 'Nothing gives man the right to claim an exception, a privilege, in the spectrum of living creatures.'

To me this seems more opinion than fact, a standpoint perhaps suitable for an ecological manifesto, a statement of principle for the Party for Animals. Not uninteresting, but to see humans as no less but also no more than the beasts hiding inside them goes too far for me. Flag-raising is an expression of the territorial urge, one I find to be a valuable insight. And in the Formula 1 driver who holds the frothing champagne bottle at waist height while rocking his hips, I do recognise something animal, or in any case something sexual.

But can all human behaviour be reduced to that? Is the Viennese waltz really nothing but a mating dance? Hardcore Darwinists believe they can knock the props out from under all human claims to uniqueness. Our dealings are prompted subtly by selfish genes, out for their own reproductive success. There is no escaping that biological predestination. How far this train of thought extends is something I once discovered in the urban jungle known as the schoolyard. One afternoon I was standing there, waiting for my daughter. Before I caught sight of her, I must have patted another child on the head. 'Aha,' said the mother beside me with a wink. 'Pet the pup and the mistress you'll chat up.'

When I let someone cut in line in front of me at the bakery, I do so only because I want to appear attractive to the opposite sex – the altruistic deed does not exist.

But still. Perhaps not every facet of human nature can be explained in bluntly biological fashion. Unnatural behaviour – that in particular – seems to me typically human. Wasting, gambling, tripping, meditating, evangelising.

To look at it from a different angle: why would people donate blood to someone they don't know? Where does compassion for the homeless or for panhandlers come from? What about the soup kitchen, the refugee centre, the rehabilitation centre? Might solidarity produce concrete advantages in the struggle for life?

'What does my compassion matter?' Nietzsche states through the character of Zarathustra. 'Is compassion not the cross to which those who love mankind are nailed?'

Compassion seems to be more an expression of softheartedness than of strength. Still, there are plenty of people who value it highly. Is that because they're so humane?

Whereas other living creatures submit to the yoke of their genetic dictates, *Homo sapiens* kicks against the pricks. We sabotage biology. Take anticonception, for example: isn't that a prime example of scuttling the evolutionary urge to pass along one's genes?

And what about homosexuality? The schoolmarm from Eijsden, face to face with the moustachioed nudists in the grass, would have had a hard time lecturing anyone about the birds and the bees.

7

E UGÈNE DUBOIS HAD HOLD of me and wouldn't let go. He lies buried on the outskirts of Venlo, along the banks of the Meuse. His daughter Eugenie had the slab on her father's grave decorated with the relief insignia of his Javanese ape-person: the skull plate and two crossed femurs.

'Two femurs?'

'It was the sculptor's idea,' John de Vos tells me. 'He thought just one wouldn't look good.'

Back in Eijsden, as soon as I hear from John about the decorations on Dubois' gravestone, I know I have to see them too. I get the chance that same week; as a professor, exam week means time off for me. My students are working on their final papers, and I have given them the following assignment:

Write a prologue to or a synopsis of the hypothetical book, as you envision it – on the basis of the material collected thus far.

In the meantime, I read the two Eugène Dubois biographies: the first one factual (written by a Dutchman and dedicated to John de Vos), the other dramatised (by an American, with an afterword by John de Vos). What intrigues me is the way that a triumph can also become a person's downfall; not in a movie or on the stage, but in real life. Dubois had succeeded at doing what no living person thought was possible. Yet he suffered so greatly under the results of his success that he withdrew from public life like a skittish and distrustful animal – overcome by paranoia.

The life story of Professor Dr Marie Eugène François Thomas Dubois (1858–1940) reads like an old tale of derring-do. As a young man he sets an insane goal for himself: to find the missing link. To achieve that, he must first break with the milieu in which he grew up. Looking back on this, he said: 'I must state that I abandoned the Catholic faith at the age of thirteen and never again submitted to its influence, neither in my feelings, nor in my actions.'

At the public secondary school in Roermond, Eugène becomes fascinated by Karl Vogt, a German zoologist who caused a public uproar with his readings, in which he mocked the Genesis account. In Aachen, rocks were thrown through the windows of Vogt's hotel by an angry crowd, while his opponents shouted: '*Haben die Affen Kirche? Haben die Affen Bibliotheken?* (Do monkeys have churches? Do monkeys have libraries?)'

Eijsden may be in a remote corner of the Dutch map, but those who grow up there speak fluent French, German and Dutch. In the language of the tourist office, the village these days lies smack dab in the middle of the 'Meuse–Rhine Euroregion'.

Eugène leaves for Amsterdam, studies medicine and passes his finals in 1884. Three years later, in the face of all expectations, he turns down an offer to become a university lecturer. Along with his wife, Anna, and their one-year-old daughter, Eugenie, he boards the *Prinses Amalia* for the East Indies as a 'medical officer' in the service of the Royal Netherlands East Indies Army.

In a rare interview given in 1928, he says: 'My life took a huge turn in 1886. My urge to trace the history of the human race surfaced once more.' The discovery that summer of two Neanderthal skulls in the Meuse Valley close to the town of Spy convinces him that this is not the 'intermediate species', not the much-talked-of missing link.

Dubois becomes caught up with Ernst Haeckel, who was sometimes called 'the evangelist of evolution'. Haeckel was the first person ever to draw up a human family tree, a leafless oak with *Homo sapiens* at its crown (on branch number 22). Two branches down, one finds the chimpanzee, the gorilla and the orangutan (number 20). The branch in between, number 21, was the one Haeckel purposely reserved for the missing link, to which he pre-emptively gave the scientific name of *Pithecanthropus* (literally, 'ape-man'). Haeckel also provided the public with a description in the form of a composite drawing of a hairy wild man with a grim expression and a listless posture.

Dubois hopes, by finding the missing link, to prove that evolution does not apply to animals alone, but equally to humans. He is a practical man, one who likes to rely on facts and experience. 'Philosophy has done little to advance the human cause,' he states. Having arrived in the Dutch East Indies, his readings and publications prompt his superiors to grant him leave for his mission. For the necessary excavation work, two army corporals will stand watch over a few dozen prisoners from nearby Fort van den Bosch. In his diary, Dubois says his Javanese forced labourers are 'as indolent as frogs in winter'. Yet they are the ones who find, in 1891, a fossilised primate molar on the banks of the Solo River, close to the settlement of Trinil.

Dubois examines and re-examines it like a man possessed. He has a couple of his own molars pulled, to place beside the find as reference material. Before long, he receives yet another fossil. It is purported to be a turtle shell, but Dubois recognises it as part of the skull of a primate. Using a dental drill, he removes the layer of baked-on, rock-hard mud. As he does so, he becomes convinced that what he holds in his hands belongs to an unknown, primal pongid: 'a successive link in the largely buried chain that connects us with the "lower" mammals'.

By then the verandah of the Dubois' house was already packed with the remains of primitive fauna from central Java, arranged by species and divided by foot-wide pathways over which no one but he himself was allowed to travel. His blindness to anything outside his mission is noticed by Anna, who accuses her husband of being indifferent even to her recent miscarriage.

During a new digging season in August 1891, after the rains stop, his coolies unearth an elongated bone. Its structure makes Dubois conclude that this did not belong to 'a tree-climber', but to a creature that walked upright. It is the femur of a '*Menschentypus*', or human-type creature, while the skullcap clearly belonged to an '*Affentypus*', or ape-type. Dubois assumes that they belonged to the same individual.

With 'perfect certainty', he states: 'The first step on our ancestor's path to becoming human must have been the attainment of an erect posture.' At one fell swoop, he adds a second conclusion to this: 'And so is provided the factual proof that the Indies served as the cradle of the human race.'

But then, in December 1892, the post brings a long-awaited chimpanzee skull he had requested from the Netherlands. By the end of the year, Dubois abandons his earlier insights. The curvature of his Javanese skullcap is much more pronounced than that of a chimp: the creature must have had twice the cranial capacity. It is not a man-ape he has found, but an ape-man (with the emphasis on 'man'). This is no word game on his part, no splitting of semantic hairs. In his notes he strikes out the original name *Anthropopithecus* (man-ape). Above it he writes *Pithecanthropus* (ape-man). With this deletion he promotes his find to the status of missing link. It is a simple slash of the pen, but one which is seen within the search for the origins of the human species as a turning point of Copernican substance.

ONLY ONE ONLINE antiquarian bookseller has a copy of the original, German-language publication:

> *Author:* Dubois, E.
> *Title:* Pithecanthropus erectus. Eine menschenähnliche Übergangsform aus Java. Batavia. Landsdrukkerij. 1894.
> *Description:* Unique copy of the actual first edition of this pioneering work in the field of human evolution, being the first clear evidence of 'the missing link', a human species considerably older and more primitive than homo sapiens, based on well-preserved fossils found near Trinil on the River Solo in eastern Central Java.
> *Price:* 15,400 euros.

That Eugène Dubois sees the ape-man as his child, dearer to him than his very own Eugenie and her younger brothers, Victor and Jean, becomes clear during the voyage back to the Netherlands in 1895. In the Indian Ocean, their steamship is caught by a storm so violent that the hull is on the point of being breached. Should there be a struggle for spots in one of the lifeboats, Eugène apparently tells Anna, you see to the children. He himself in that case will apply his survival skills to rescuing his wooden crate with the skull, femur and molar.

Upon his homecoming, Eugène Dubois is counting on an homage and a royal decoration. Or, at the very least, a curtain call, applause. With his ape-man remains, he has provided empirical evidence to support Darwin's theory. Yet humankind is unimpressed, as is his own mother. At the parental home in Eijsden, she glances quickly at the bones he has brought back from Java. 'What good are they to anyone?' she apparently snaps at her son.

His father, who would have understood – who would have been proud of him – has by this time already died.

In the middle of that same week, I find myself at the graveside of Eugène Dubois. The sepulchre is mossy, covered in green smudges. It stands there in cold light filtered through the branches of a maple. Following John de Vos's instructions, I walked past the gravestones with crosses to the field for stillborn (unbaptised) babies and atheists. Grave number NH226 BR is on unhallowed ground. On the covering slab I recognise the relief of a skull with two crossed bones. Anyone wishing to be sure that this is not the final resting place of some pirate captain will have to look closely to make out the contours of the Javanese ape-man in the carved skull – with his prominent brow ridge seen from above.

Despite my prior knowledge, the symbolism touches me. During the first half of his life, Eugène Dubois had luck on his side. In the struggle he took on, he maintained the upper hand for a long time, but ultimately – or so speaks the ornamentation on his grave – his trophies turned the tables on him.

During class, we had paused to consider the antonyms BURY and DIG UP. We asked ourselves: who among the dead do we leave in peace, and who do we not?

In all cultures, the desecration of graves and corpses is a punishable offence. Even archaeologists and paleoanthropologists themselves, one supposes, would like to be left in peace after they die. Still, it is their job to exhume human remains.

No one was bothered by the exhumation of Java Man or Flo. The same went for the illustrious Neanderthal of Spy, found in 1886 in what looked like a grave. That the deceased had been laid to rest in ritual fashion apparently formed no obstacle to digging them up again. Human curiosity won hands down over deference and piety.

UPON HIS RETURN to Europe, Eugène Dubois' rocky road began. With the exception of a few scholars like Ernst Haeckel, most of his contemporaries refused to see Dubois' fossils as those of 'a creature on the threshold of becoming human'. They came up with only curt comments and criticism.

Dubois, they said, had done sloppy work. The description of his discovery took up a mere nineteen pages, which was unscientifically terse! The supposition that the femur belonged to the same creature as the skullcap was nothing more than that: a supposition.

It was found fifteen metres further along. Fifteen metres and how many centimetres?

In the same layer of earth. Dubois could claim that was so, but could he prove it?

'This monkey theory is marked by monkey business,' an instructor at the Ethnology Museum in Leiden wrote in 1895. Dr Dubois had 'put science to shame' by presenting his so-called missing link in such painstaking detail, 'as though he had known him personally'.

Comments like that wounded Dubois deeply. He became difficult, for himself and for his loved ones, almost tyrannical. Every morning he made his young sons, Jean and Victor, take a cold bath, after which they had to perform stretching and endurance exercises on the gymnastics equipment their father had built.

At first he defended himself, during readings in Leiden, Brussels and Berlin, but soon enough he started to shun the public arena – and he took his fossils with him. Although they rightfully belonged to the Dutch government, as the financier of the excavation work, Eugène Dubois hid them under the floorboards of his house in Haarlem. When anyone rang the bell, he sent a maid to the door to say that the professor was not at home. Upon being turned away for the second time, an American scholar from the Smithsonian Institution in Washington left behind his calling card with a handwritten message on the back: 'The anthropologists of the world have a great deal to thank you for, but it is a damned shame that one is not able for scientific purposes to even glance at the specimens.'

When Dubois was asked to lend his ape-man remains for the exhibit at the World's Fair in Paris, he refused.

In an article from October 1900 titled 'How the Pithecanthropus Made It to the Trocadero', found in the Delpher databank, the Paris correspondent for the *Algemeen Handelsblad* stood up for Dubois. Even though he did not let them use the original bones, Dubois did go to the trouble of making a plaster statue with his eleven-year-old son, Jean, as his model: an artist's impression of the Javanese ape-man. The man-sized form – bolt upright, with sagging shoulders and gawkish arms – held a set of antlers and gazed at them in astonishment, as though thinking about how he might use them to some end.

The sculpture was put on public view in the pavilion dedicated to the Dutch East Indies, at the foot of the brand-new Eiffel Tower. 'It is of the essence that the glory be preserved for the Netherlands, now that competitors are entering the lists against her,' the *Handelsblad* correspondent wrote. He ended with a call for the Dutch government to finance more excavations on Java.

ON THE SLAB OF EUGÈNE DUBOIS' grave lies a half-withered floral arrangement of holly and pine branch, arranged artfully around a chunk of tree bark. Someone placed this here, and not very long ago. Consulting my phone, I discover where the bouquet came from: 'Homage to Dubois on 75th Anniversary of His Death'. The *Omroepvenlo.nl* site makes mention of the laying of a 'funeral wreath by 92-year-old Mrs Truus Geerlings, who as a young girl worked as a maid in the professor's home'. The bark and branches of holly come from De Bedelaar, the estate where Eugène Dubois spent his final years, and where he died of a heart attack on 16 December 1940.

The organisers of the homage, the website says, were nature lovers from the area around De Bedelaar. I find a phone number, and within two hours I'm drinking a cup of coffee in the home of Sjra van Horne, an amiable former alderman from the municipality of Haelen, under which De Bedelaar falls.

'Sjra?'

'That's Limburgish for Gerard.'

Mrs Geerlings is his aunt. 'I'm afraid there's not much point in arranging a meeting,' he says. 'Sometimes she's completely lost. Then you get to hear the same story over and over again, the one about the bicycle.'

What, I ask him, is the story about the bike?

With undertones of endearment and pity, Sjra tells me: the professor used to send Auntie Truus, whom Dubois called 'Truudje', to the village post office every day. 'There were always big piles of letters for him, in envelopes three to five inches thick, she always told us. When old Dubois found out that she borrowed her mother's bicycle to run the errands, he immediately bought a bike for her, a new one.' Sjra van Horne smiled benignly. It was another anecdote, though, that made more of an impression on him. 'On occasion there were packages for him containing amputated human limbs, from the hospital. They were always packed in wax paper. When it was hot out, the packages would start dripping on the way home.' Dubois' gardeners had the task of stripping the meat from the bones, and bringing them to the professor. When little children came to visit, which occurred only rarely, Eugène Dubois would wrap a measuring tape around their head and jot down their age, gender and skull circumference. He lived with a monkey.

Since his retirement, Sjra has spoken to all of Dubois' living acquaintances. 'The professor was an eccentric fellow, you know, who did eccentric things.'

'Such as?'

'The maids sometimes saw him squatting over the plants in his vegetable garden. He fertilised them himself,' Sjra says. 'Tomatoes. Almost no one had tomatoes in the garden back then.'

Dubois' American biographer even referred to him as an 'elderly lecher' and 'paranoid'. Van Horne is not surprised at all. 'He was a menace to all the farmgirls.' The hired girl Caris came running into Haelen one day, weeping hysterically. The gentleman of the house had locked her in a room, but she had escaped through the roof. Dubois, he tells me, lived separately from his wife, Anna; they each had their own wing at De Bedelaar.

'There was a son of his, in the village. A bastard son. Everyone knew about it.'

WITH SJRA VAN HORNE's hand-drawn map clutched to the wheel, I drive my rental car to De Bedelaar. 'Professor Dubois Lane 2' reads the road sign next to the drive. Followed by 'Authorised Persons Only'. Across a moist blanket of autumn leaves and beechnut shells, I make my way to the white-stuccoed house. In the woods to my right a brick archway serves as entrance to an icehouse; in the winter Dubois had blocks of ice sawn from the nearby fen, which served to keep his vegetables cool in summer. De Bedelaar once comprised forty hectares of forest, thicket, heath and open water. The lord of the manor had tried to transform it all into a prehistoric park. Dubois hoped to imitate in the present the natural conditions that prevailed during the 'Tiglian' stage of the Pliocene, the era in which the river clay of Tegelen was deposited. To that end he lowered the water level in the fen, set out a thousand rock-bass fry and six hundred tench, and planted rare tree varieties, including swamp cypress and the *Sequoia gigantea*, the forest giant of the American west.

But no matter how deeply he withdrew into his own nature reserve, Dubois went on worrying about his fossil collection – including the skullcap, femur and molar – which he had at last bequeathed to Leiden University. More than once he sent a warning to the rector for the Dutch bishops, who he was sure wished to steal and destroy his evidence against Creation. When the fossils were moved from one depot to another, he forced the school to arrange a police escort, and when Dubois heard that a professor of archaeology wanted to inspect the contents of the strongbox, specially designed for the purpose by the Lips lock company, he did his best to prevent it, arguing that the professor in question was a practising Catholic.

Dubois almost never left his estate, but that did not keep him from taking swipes at anyone who deigned to follow in his footsteps. It was as though he were claiming the entire domain of primal anthropology all for himself. The more imitators came along in the hunt for human skulls, and the more discoveries this new generation brought to light, the more precipitate Dubois' attacks became. It was Ralph von Koenigswald – who had uncovered two skulls on Java that so closely resembled Dubois' ape-man that he named them Pithecanthropus II and III (the celebrated 'Sangiran skulls') – who served as the special focus of his ire. The newly found skulls also came from old layers of riverbed along the Solo, but a few villages upstream from Trinil.

In his memoirs, Von Koenigswald noted: 'I thought Dubois would be overjoyed, but I was sorely mistaken.' The two had apparently met once in the flesh, in Haarlem in 1937, where (after a failed attempt on Dubois' part to duck the appointment) they had a cup of tea together. Von Koenigswald described him as a 'broad-shouldered man with a stereotypical, almost timid smile'.

When Von Koenigswald's publication – *Neue Pithecanthropus Funde 1936–38* – rolled from the presses, Dubois became riled. From his study at De Bedelaar he espied a new conspiracy, a pointed attempt to relegate his Javanese ape-man to the shadows. Von Koenigswald was in cahoots with Franz Weidenreich, a German working in China who was causing a furore with a fossil skull he was calling 'Peking Man'. By describing their fossils as hominin ('more man than ape') and then pointing out the similarities with Dubois' find, the two drew the conclusion that the Javanese ape-man was not the missing link, but 'merely' a primitive form of human. In analogy with the name 'Peking Man', they recommended that Dubois' ape-man from now on be referred to as 'Java Man'.

Paleoanthropology was in the early stages of development and becoming increasingly professional, but Dubois stuck to his original idea. He and no one else had 'bridged the gap between human and animal'. In order to defend the status of his pithecanthropus as the missing link, he launched – in 1940, the final year of his life – a tripartite assault on Von Koenigswald's integrity, including the implication that Pithecanthropus II and III were fakes. His rival's working method was faulty, Dubois claimed, because he paid local diggers by the fossil. If they found a backbone, they would first break it into pieces in order to receive multiple rewards. So why couldn't the local population just as easily have satisfied his hunger for fossils with intentionally distorted skulls?

Von Koenigswald replied in a letter from Java. 'As I wrote to you already, most highly respected professor, the latest discoveries have not served to confirm your opinions.' (The observation that Dubois 'reacted unreasonably, like a lover scorned', was one Von Koenigswald saved for his memoirs.)

By the end of his life, the only person who could still stand Dubois was his daughter, Eugenie. 'I am considered hostile towards others,' he complained to her. 'My opponents do not understand me, or rather: they do not wish to understand me.'

Even as a squadron of German bombers flew high over De Bedelaar on its way to Rotterdam, he remained obsessed with 'the other party'. 'The longer this continues, the clearer it becomes to me that they are not interested in the truth.'

Only some six months after the start of the occupation, in November 1940, was he capable of deeper reflection. The realisation dawned on him that 'the ideological principles of the current world war' could be traced back to 'anthropological issues'. Back to exactly those same issues in which he had involved himself: quarrels over the shape of skulls, anomalous anatomies, and above all the ordering of human types by reference to superiority and inferiority.

DE BEDELAAR, AS it turns out, was renovated and refurbished in 2017. The villa has been split up into twenty-five rooms, lived in by an equal number of men: men in need, who cannot take care of themselves because they are mentally ill or at the end of their tether.

Sjra van Horne had referred to De Bedelaar as a home for men at odds with themselves, as an open mental institution. In the distance I can see residents behind the windows of the sun porch. One is walking around outside, close to the bat house; another is pushing a wheelbarrow filled with sand for the cement mixer. In my imagination I see them as descendants of the former owner. As though the spirit of Eugène Dubois, multiplied by twenty-five, still walks this estate.

8

TO GET A HANDLE on the swarm of extinct hominids, I decide we should hang a poster of the human family tree in Classroom 0.04. What, according to the latest scientific insights, does that tree look like? Just as a professor of surgery has recourse to a dummy with removable organs, I need a steadier frame of reference when it comes to the genealogy of our forebears and kin. Along which branch might one find the Neanderthal, and on which branch *Homo erectus*?

The hitherto-invisible Freek, a quiet new media student, is given the task of sourcing the human family tree. But soon he hands back the assignment. What we are looking for doesn't exist, he says. 'I found a lot of them, actually, but they're all different.' There are, as it turns out, as many family trees as there are finders of special skulls.

Freek lays a whole stack of human family trees on the table. The older versions, from before World War II, all have their roots in Asia. In imitation of Ernst Haeckel, Dubois drew one too, one in which his pithecanthropus fills the gap between a fossil, chimpanzee-like creature found in Pakistan and *Homo sapiens*.

'I am cognisant of the extraordinary temporality of such family trees,' Dubois wrote. 'But I also know that parts of them sprout, and from those new life blossoms forth.'

A lovely turn of phrase, but little had remained of Dubois' own family tree. His successor, Von Koenigswald, considered the drafting of such genealogical trees to be an indispensable aid. They reminded him of the 'skull trees' he had seen among the Papuans: real trees along the paths leading into a village, their branches decked out with the severed and dried heads of local enemies.

Even though Von Koenigswald did not consider his own sketches the be-all and end-all, he was rigorous in pruning away at the ape-man branch. Dubois' pithecanthropus was finally awarded a spot as the holotype of *Homo erectus*: not half-human, half-beast, but very much our direct ancestor. He ultimately adopted only one of Dubois' views: the genealogical trunk belonged in the Dutch East Indies. Between 1890 and 1940, paleoanthropology developed thanks to Dutch scientists who, despite their disunity, situated the cradle of humanity in Asia.

In the genealogies from 1950 on, however, Africa took the place of Asia. Unlike Wallace, Darwin had always affirmed that we had lost our fur and tails in Africa. Not only did the post-war fossil hunters shift their hunting grounds to the African savanna, they transplanted the tree as well. The first to occupy the recent vacancy for the missing link was the Taung Child (whose head was no larger than an orange). After that, a 1.6-million-year-old primitive stonecutter from East Africa, discovered in 1964, became the first primate in the genus *Homo*. Based on these finds, the 'out of Africa' theory held sway for decades: a conception that lends itself to illustration in the form of a weeping willow, rooted in southern Africa and branching out by way of Siberia, with hanging twigs reaching through Malaysia all the way to Java. But this tree too was felled at last: the snarl of bifurcations that is left is referred to today as 'the family bush'.

Each new discovery seemed to provide its finder with an instrument honed sharp as an axe. The striking thing was the gusto with which the paleoanthropologists sank that axe into each other's work. Sooner or later, their egos or their patriotism got the best of them. Accordingly, we stumbled upon a Georgian whose team had unearthed five hominid skulls from a lava pit along the Silk Road. With every trophy they secured, he raised the same toast: 'In human terms, I should now be sad. But I am extremely happy with this dead person. What pleases me most is that he died here with us, and not a few kilometres further in Armenia.'

In fact, I suggest to my students, we should really be drawing a very different family tree: that of the finders. It is a casual remark, a dropped phrase. One week later, however, Freek comes back with two sheets of A4 paper. Sketched on the first is a sort of bush; you can't really call it a tree. The plant climbs up along the Y-axis; a vertical timeline that began two million years ago (with the rise of *Homo habilis*) and reaches all the way to the present (*Homo sapiens*). Half-buried at the roots is Lucy the ape-woman and the Taung Child as non-human transitional forms. The longest twig on the graph is occupied by gawky *Homo erectus*.

The second sheet of paper is in fact an adaptation. Instead of two million years, the timeline covers only two hundred years. Along the branches of the bush, Freek has replaced the names of the hominid types with those of a few notorious hominid hunters.

The place of Lucy and the Taung Child at the roots is now taken by CHARLES DARWIN (1809–1882) and ALFRED RUSSEL WALLACE (1823–1913).

Along the branches themselves, from the bottom up:

EUGÉNE DUBOIS (1858–1940, finder of Pithecanthropus, a.k.a. *Homo erectus*)

RALPH VON KOENIGSWALD (1902–1982, finder of Pithecanthropus II and III, a.k.a. *Homo erectus*)

THEODOR VERHOEVEN (1907–1990, finder of the proto-Negrito of Liang Toge, predictor of the existence of a human avatar on Flores)

PAUL SONDAAR (1934–2003, unsuccessful searcher for this early Flores Man)

TEUKU JACOB (1929–2007, finder of *Homo erectus*, denier of the existence of *Homo floresiensis*)

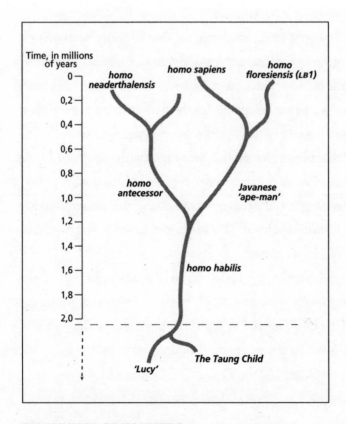

FAMILY TREE OF HOMINIDS

MIKE MORWOOD (1950–2013, finder of LB1/*Homo floresiensis*)

To achieve a mirror-image effect, I put the two sheets of paper up on the board, side by side, with magnets.

The found fossils, side by side with their finders. The former remain silent, regardless of whether they themselves possessed the gift of speech. The latter are garrulous. The question is what we are to do with all this.

In a book, a family tree at the start suggests an epic tale. Take *One Hundred Years of Solitude*, for example. In the very first sentence we make the acquaintance of Colonel Buendía, who, standing before the firing squad, thinks back on 'that distant afternoon when his

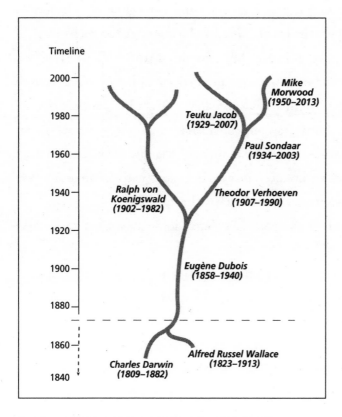

FAMILY TREE OF PALEOANTHROPOLOGISTS

father took him to discover ice'. At the end, Gabriel García Márquez has the final living descendant of the Buendía line, along with the mythical village of Macondo, wiped from the face of the earth by a hurricane. And between the opening and the ending, the book teems with family feuds and intrigues.

All stories are served by struggle, between fathers and sons, mothers and daughters(-in-law). But within the journalistic genre of reportage, you can't simply dream up heroes and anti-heroes. One must make do with an existing cast of characters: a handful of hominids and their discoverers. The latter may not have been related by blood, but they are certainly connected in strife.

Freek's family-tree adaptation is more than a gimmick. The successive graverobbers had kept a sharp and cunning eye on each other. As soon as they got the chance, they committed 'patricide'. Now I suggest that we have our imaginary camera pan from the hollow-eyed skulls to the tanned faces of their finders. Can we, on the basis of their characters and behaviour, draw any conclusions about the idiosyncrasies of modern humans? Might Freek's alternative family tree help us do that?

My students pounce on these questions with carnivorous glee, and tear them to pieces, as they have been taught.

'Family tree?' someone says. 'Why do we use the image of a tree, anyway?'

Their knock-down methodology is enough to drive you mad at times, but this is a crucial question. I know I have to play the ball back into their court. 'Right, what does that say about us?'

We decide at last that the trunk stands for constancy, and the tree itself for life. And that the tree has its roots in the fertile soil, to which we all return in the end. Like a supra-organism, the family tree comprises an abundance of individuals. Whether you're talking about a royal dynasty or simply some family chosen at random, the relations consist of vital juices, blood ties, inherited DNA. Everything revolves around consanguinity and, accordingly, reproduction and sexuality.

This latter conclusion leads us onto a new track. Someone mentions the Mormons. After all, if there is one group – alongside the paleoanthropologists – that is involved in systematically unravelling the human family tree, it's the Mormons. The Church of Jesus Christ of Latter-day Saints has the objective of placing all of humankind, with no exceptions, the living and the dead, onto a family tree. That is to say, onto *the* family tree.

Whereas the cranium-crackers begin with the oldest hominid fossils, the Mormons work in the opposite direction, starting with the living, in the present. Based on the estimate that some seventy billion people have walked the earth from the very beginning to the present day, they try to gather the personal particulars of every earthling. From their temple complex in Salt Lake City, the Mormon FamilySearch.org company buys public registers from as many countries as it can, and links them together. The company logo is a tree with fresh, green leaves. Whatever you might think of them, the Mormons exhibit a termite-like assiduousness in gathering their genealogical data. With a knack for publicity, they actively cultivate enthusiasm for TV shows with names like *Who Do You Think You Are?* One of us recalls an episode in which Cindy Crawford – 'a girl from the Midwest' – was told that she is a descendant of Charlemagne.

Stunts like these draw attention away from one of the most notorious quirks of the Latter-day Saints: their polygamy. Male church members, or at least the top dogs among them, maintain their own harems, and often sire dozens of descendants over the course of their lifetime.

After brainstorming, we decide that the Mormons combine the procreative urge with research into lineage. This combination of polygamy and genealogy may seem like a digression, but I invite one of the students to wander down this side track even further. Literature major Elfrieda raises her hand to volunteer. A friend of hers has been on an excursion to the genealogical treasure-trove of the Mormon Church, the Granite Mountain Records Vault: a databank in Utah built – in true science-fiction fashion – in an underground maze of corridors at the heart of a gigantic rock.

The safe full of ancestral data, it seems, actually exists. We have discovered a new cave.

WHEN IT COMES to our own research into the family tree of the paleoanthropologists, we are making steady progress, marked by leaps forward from time to time. The successive hominid seekers have in fact commented publicly on each other's work.

'All prehistoric research is subject to criticism after the fact,' Father Verhoeven wrote. 'One needs only recall Dubois' work in Indonesia.'

Mike Morwood, himself the object of Teuku Jacob's censure, had dismissed Paul Sondaar's work in a footnote as mere earthmoving. The victim of this attack struck back by publicly accusing Morwood of 'knowingly' co-opting his 'primitive-man-could-already-sail' theory: 'There was a chance of international recognition, and then someone comes along and does this.'

In a festschrift collected in honour of Sondaar's sixty-fifth birthday, we discover a peculiar 'family photo' from 1970. In the foreground, sprawling on a two-person settee, is Ralph von Koenigswald, with fleshy cheeks and a neat wave in his hair. Beside him his startled-looking wife, Luitgarde. Standing behind them are three doctoral students, including a youthful, bearded Paul Sondaar. Like witnesses at a baptism, they look on as their mentor cradles one of his fossil skulls in his arms. Von Koenigswald would have had every reason to be satisfied, were it not that the Indonesian government – with Teuku Jacob as its adviser – had declared him persona non grata in the late 1960s. His PhD supervisor Von Koenigswald was no longer welcome on Java, not unless he came in person to turn in his clandestinely exported fossils to customs in Jakarta.

It all seems like the product of some screenwriter's imagination: we are dealing here with a cast of pushy antagonists, and with protagonists who push right back. Freek's parallel family tree presents a structure with epic features. In order to understand the rivalry between our characters, though, we also have to appreciate their hidden agendas. 'What makes Sammy run?' is the classic textbook reference to a character's motives. That can often be expressed in a single word. Illwill. Greed. Pride.

Next to the name Dubois, we write 'assertiveness'. He possessed the authoritarian traits and the madness of a romantic. The Don Quixote type. The others, too, resemble classic characters, ambitious and easily offended. Except for Father Verhoeven. He is the exception, even in the way he dressed, with his cassock and clerical collar. Why did a cleric like Verhoeven go rooting around in the earth? We have no idea.

The key, in rather cryptic terms, is handed to us by someone who had known him well: 'You people should remember that from the age of twelve on, Verhoeven had no women in his life.' Gert Knepper, a man in his sixties who calls himself a 'New Testament scholar', is the one who points this out to us. He knows a lot about Father Verhoeven, but doesn't really want to tell us anything. At our first contact, he actually seems shocked, as though he would rather not have been identified as an expert on Verhoeven.

Mariëlle is the one who tracked him down, via a Russian blog where he swapped observations with someone called Michail Tsyganov. Things like the scan of an undated picture postcard showing a Garuda Indonesia airlines DC-8, on which 'Th. Verhoeven' sent his regards to Gert Knepper. Mariëlle ran the Russian dialogues through Google Translate and called Gert at home one evening.

'Verhoeven was his high-school Greek teacher in 1969/1970,' we read the next morning in our Dropbox. 'Verhoeven's broad Brabant accent with Indonesian overtones was an attraction. The students kept trying to get him to talk about Flores.'

Their teacher had limped as he walked back and forth in front of the blackboard. 'It's because of the accident,' he would explain. What had made a major impression on Gert was how easy it was to get a driver's licence on Flores. 'They loan you a car, and if you bring it back in one piece within the hour, you've passed your exam.'

Gert and his teacher became friends. For years he went on visiting Verhoeven – and Paula too, later on – at his home. Sometimes they went out together for an afternoon in Verhoeven's yellow Opel, looking for hand axes in the Meuse Valley.

'Verhoeven left him his Greek textbooks, he has them at home,' Mariëlle noted. 'Gert himself also became a Greek teacher later on.'

Gert Knepper, as it turns out, has written a biography of Verhoeven which had not been published. He keeps the text on two thumb drives. His project had been prompted by the news of the discovery of Flores Man on 28 October 2004. In the international news reports, Gert had occasionally come across the name of his former Greek teacher, praise for the pioneering work of the 'Dutch missionary'. To Gert's irritation, though, this was always followed by the remark: 'He married his secretary and returned to Europe.' Five years later, in 2009, Gert told Tsyganov the blogger: 'This year, or maybe next year, I'm going to publish my biography of Verhoeven.'

That never happened. Gert can't really explain what went wrong. Not even when Mariëlle and I meet up with him at a lunchroom in Leiden. The man we meet there is meticulous – close-cropped hair, leather bag on a shoulder strap. He speaks freely and to the point. From his bag he produces no manuscript, not even a thumb drive. He has no desire to give up the subject of his biography to us just like that. What is needed is a fresh rendezvous, on more neutral territory, to help Gert come around.

The next time, we meet on the main concourse of Utrecht's central train station. Mariëlle and I present him with the facts as we know them, such as Theodor Verhoeven's part in studying the charred, rolled-up bodies from Pompeii's 'Garden of the Fugitives' in 1946. Once Gert changes his mind, there is no need for us to prompt him any further.

'After his mother's death, his father started drinking,' he tells us. 'Theo escaped to the preparatory seminary, but that was an exclusively male community. A boys' boarding school.'

Seated there at a table beside the glass wall of Bistro Central, with a bird's-eye view of the muffled swarm of commuters, Gert warms to his subject. One by one, he brings up the scandals by which the Catholic Church has recently been sullied.

I interrupt him. 'Are you trying to say that Father Verhoeven was abused as a boy?'

'Exactly,' Gert says. 'At the preparatory seminary, by one of the brothers who taught there.'

We have just made the switch from coffee to beer. Pursuing him with questions feels awkward, but also inevitable. 'What are you saying?'

'He was raped.'

'Is that what you call it, or did he use that word himself?'

'What do you want me to say? Penetration?'

At home Gert has a twenty-page letter in which Theo Verhoeven looked back on his life. It describes 'expressively' the depravities that the brother with his rolled-up cassock performed on him. 'It may sound strange to you two,' Gert says, 'but in that letter he absolves the man of all guilt and puts the blame on the church's vow of celibacy.'

Hidden behind the glory of Verhoeven's archaeological achievements was an abyss, a yearning. Not for recognition, as it was for Dubois, but for intimacy. 'Towards women,' Gert clarifies. As a priest, Verhoeven was always obliged to maintain at least an arm's length of distance from women and girls. Only during the sacrament of baptism was there a moment of physical contact, but in that situation the gesture was purely sacral. Frustration led Theo Verhoeven to make jokes about it. On Flores, it was the custom for mothers to keep their child on the breast during the baptismal rite, and his role was to sprinkle the correct bump no less than three times, in the name of the Father, of the Son and of the Holy Ghost. 'Sometimes you miscalculated.'

In those twenty pages Gert had inherited, typed by Paula, Verhoeven condemned celibacy as 'unnatural'.

'It is,' I concur.

'It's unhealthy,' says Mariëlle.

Gert tells us that there were periods when Theo longed for a woman so badly that he became depressed, almost suicidal. 'When that happened, he started behaving recklessly.'

We ask him to give us an example. Gert hesitates. Then he says that, during World War II, Theo had helped Jewish children from the city go into hiding in Brabant province. 'Thirty,' was all he himself ever said about it. Verhoeven had never boasted about saving those children, but Gert knows that there was something more behind those acts of heroism. According to him, the risks Verhoeven took were a sign of contempt for his own safety – as though his personal chances of survival had become irrelevant. If it was heroism, then it was heroism born of despondency, for which Verhoeven himself blamed the constricts of celibacy.

'"Celibacy made my life a living hell," that's what he wrote, literally,' Gert says.

It feels as though we have reached the bottom, the lowest point of Verhoeven's life. 'Bedrock' is what those who excavate for a living call this; below it there is nothing.

We sip at our beers. Below us on the concourse, other travellers are walking by; still, after one whole hour, the look of the human swarm has remained unchanged. The commotion down below has a somewhat calming effect.

Gert empties his glass and puts it on the table. 'At the seminary,' he says, 'Theo fell in love with a classmate.'

Mariëlle and I are both startled; we haven't expected any further revelations.

Gert tells us that Verhoeven, in the absence of women and 'on the rebound', started feeling attracted to boys. In accordance with the mores preached by the brothers who had come to the dorms in the evening to molest him and his classmates, the future discoverer of the prehistoric dwarf elephant and the giant rats of Flores realised how base and sinful his feelings were. It confused him.

UNLIKE THE ANIMALS, they say, humans can experience shame. They cover their nakedness. High on our steadily growing list of 'most commonly noted differences between humans and animals' is the entry 'wears clothing / no clothing'. Linked to this, during the Victorian era, was the notion that animals were promiscuous, while people, or at least the most respectable among them, were monogamous. Alongside this moral distinction, there was also the lack of a periodic (compulsive) mating season among *Homo sapiens*. The cat is on heat, the bull is ruttish, but we humans no longer let ourselves be driven by our lusts. By dressing in bearskins, it was said, our hominid ancestors had risen above the animal state of shameless nudity.

In the Protestant slough from which I hale, shame and guilt went together like a horse and cart. Guilt is not the right word, though: sin was what it was all about. Guilt multiplied by disgrace. Anyone who let themselves be tempted as easily as Eve, and later Adam, was bound to rip a leaf off the fig tree out of sheer humiliation.

Most of the virtues I was taught as a teenager are, in fact, inhibitions. Cutting ahead in line? The first shall be last. A dog gulps, a cat laps at its milk, but you hold yourself in check. Civilisation stands or falls on how well you bridle the animal inside yourself. Gluttony is a lack of restraint, alcohol makes one licentious, sex before marriage is fundamentally wrong.

I can still summon the dismay I felt the time – when I got back from my first vacation without my parents – I shook out my sleeping bag in the garden in the presence of my mother and a pair of panties fell out of it.

My mother bent over them like a stork, shivering with disapproval.

'Oh yeah, that's right,' I said hoarsely. 'I loaned my sleeping bag to some Danish girl. She was cold.'

My mother didn't believe me, but her will to believe was stronger. She picked up the panties and tossed them in the compost bucket.

Thinking back on that moment of shame, I realise how closely it fits the person we were talking about. Everywhere, reproduction is subverted into unnatural channels by a host of rules, written and unwritten. Whether it's China's one-child policy or the Indian practice of abandoning baby girls (or 'selectively aborting' them as foetuses), we supplant Darwin's natural selection with an artificial one. By means of contraception, abortion, sperm donorship, embryo selection and a whole gamut of taboos, we divert the course of evolution. Nature is allowed to run its course in many areas, but in every culture, when it comes to reproduction, morality keeps watch.

As I write this, the morning news on any number of websites has been showing a video of two Indonesian men on a gallows. 'Public shaming in Aceh, Indonesia,' the BBC reports. A screen pops up to ask whether I am sixteen years of age or older. After tapping 'play', one hears a barrage of raw shouts. 'Hit him,' the male crowd chants. 'Harder, harder!'

I see a masked guard. He has a brown hood over his head, with yellow stitching and two holes cut out for the eyes. He holds a cane, and as the crowd cheers he brings it down on the backs of the two scapegoats. Trembling in shame and pain, they submit to their beating. Across the screen rolls the text:

THE MEN'S CRIME?

GAY SEX.

Vigilant neighbourhood watchmen had forced their way into the men's room, finding them naked in bed together. By having sex together, the men were refusing to cooperate and pass on their parents' genes in the proper way. The public prosecutor called for eighty strokes of the cane each, but the judge raised the sentence to eighty-three.

I would like to have discussed the video with the students, but classes are over now. It would have been interesting to contrast the anti-gay witch hunt in Banda Aceh with the poster campaign run in Rotterdam in the spring of 2017. In one of the pictures, a young Muslim woman, recognisable by her headscarf, is kissing a boy whose yarmulke tells us he is Jewish. In the background: the steel moorings of the Erasmus Bridge. Another poster in the same campaign showed two women kissing. They were also standing beside the iconic bridge that bears Erasmus's name. The message, edifying and utopian but unmistakable: *This is normal in Rotterdam.*

What I would like to have asked the students is: does Rotterdam enjoy a higher degree of civilisation than Banda Aceh? Are 'we' in Holland more humane than 'they' in Sumatra?

ELFRIEDA, WHO WILL ultimately receive an A minus from me, has put a premature halt to her investigations into the Mormon Church. She'd started by entering her name into the network of FamilySearch.org. Before long, she had wandered into the labyrinth of her own descent. This stroll past generations of her own ancestors was exciting, but also had a dark side, Elfrieda reported: 'My privacy.'

The deeper she delved into her past, the more cunningly the makers of the family tree tried to lure her in. If she sent them a little sample of saliva, she was told, the supercomputers in the rock outside Salt Lake City could shed light on even more branches of kinship.

FamilySearch.org worked with Ancestry.com. From the latter
you could order a kit with a little spatula you used to scrape the inside
of your cheek, and then you'd put it, wet with saliva and mucus, into
a tube with a screwcap. In late 2016, Ancestry.com had DNA samples
from four million earthlings. The fine print, however, had a totalitar-
ian ring to it. Anyone sending in mucus was granting carte blanche
for the 'analysis, processing, transfer, distribution, communication
and exhibition' of the owner's DNA profile. The Mormons claimed
for themselves the right to use this mucus for the purposes of 'per-
sonalized products and services' (for example: customised Facebook
ads in the event of a likely aptitude for musicality). The trade in this
was unlimited, worldwide, and not subject to any claims from the
donor. The duration of the contract: perpetual.

Although Elfrieda had given away nothing but the IP address of
her laptop, her Mormon hunt soon rounded on her: the Mormons
began pursuing her. 'Dear Elfrieda,' began the email sent to her 'on
behalf of Kimberley, Jacqueline, Monique, Michael and Yentl'. The
five of them thanked her profusely for her interest in the Book of
Mormon, the sacred text that church father Joseph Smith had had
dictated to him between 1820 and 1827. 'The missionaries in your
town will probably come by and bring you the Book of Mormon this
week or next week.'

Even for a class that would net her five full credits, Elfrieda had
no desire for a visit from the Mormons. By way of compensation, she
had made a careful study of the ultramodern ancestral cave north of
Salt Lake City (where it is always 15.5 degrees Celsius, with 45 per
cent relative humidity), as well as of Mormon mores.

As far as the bunker went: between 1960 and 1966, it had been
excavated in the 200-metre-high granite wall of a valley left behind

by receding glaciers. As evidenced by the double set of front doors, weighing 9 and 14 tons apiece, the Mormons took into account a possibly apocalyptic end to life on earth – after which only their registers of births, deaths and marriages would remain.

The Mormons' polygamy, as it turned out, also had futuristic traits to it. On 12 July 1843, polygamous marriage was revealed to Joseph Smith as a doctrinal tenet of his church-in-the-making. A devout man was allowed to establish sexual alliances with an unlimited number of women. One year later, when Smith was lynched for heresy by an angry crowd in Illinois, he left behind thirty widows. The record holder, Warren Jeffs (who has been serving a life sentence for statutory rape since 2011), is said to have been married to seventy-six women and (underage) girls. According to Mormon dogma, the human family tree was not only to be charted, but also to go on growing and blossoming lushly – in order to baptise as many specimens of *Homo sapiens* as possible, both the living and the dead.

POLYGAMY AND CELIBACY seem like polar opposites. Two lines of action, diametrically opposed. In my mind's eye I see, in the shelter of a rock wall in Utah, harems where the women are pregnant more often than not. At the same time, I can't help but think of the pink sisters of Steyl, kneeling devotedly in their chaste habits.

Maybe we are dealing with extremes here – but isn't that also something that characterises us as a species?

At fifty-nine, Father Verhoeven had buckled under the burden of celibacy. As a young man he had chosen an asexual life, devoid of physical contact. Solemnly, with his hand on the Bible, he had taken the vows of penury, obedience and chastity. Distancing himself from the world's possessions was no problem for him, submitting to decisions from on high without grumbling was hard, but celibacy simply crushed him. It was a cross he could not bear all the way to the end.

There was no direct connection between the celibate life and Theo Verhoeven's decision to spend eighteen of his best years searching for fossils, Gert tells us. Indirectly, he feels, a connection did exist: Verhoeven was in search of distraction and needed to put his passions and energy into something. Upturning the soil alongside dedicated pupils helped. Father Verhoeven drew a satisfaction from it that he found nowhere else. But at one swoop – and a fell one at that – all this was taken from him. One afternoon, in the mountains overlooking the Flores Sea, he drove into a ravine; his mission jeep was dashed against the rocks a hundred metres below. Verhoeven himself was thrown from the vehicle; in his cassock he rolled down the slope, coming to a halt against a boulder ten metres below the road.

The last hours of 1966 he spent in Flores, hooked to a drip. There were moments when it seemed he would tumble over the edge. The next year was taken up with repatriation and rehabilitation, after which he miraculously experienced his own resurrection.

Many survivors of near-fatal accidents suddenly believe in God or make solemn promises to live more devoutly. Theo Verhoeven, however, did the opposite. Once back on his feet again, hesitantly at first and on crutches, he renounced the priesthood. The abbot of the Society of the Divine Word wanted nothing more to do with him, especially after he married a former nun in the early 1970s.

How Theodor and Paula found each other is a question that drifts over our table for a few seconds. We have already paid the bill and are skimming the departure times for the next few trains home.

'Through an advert in the personals column,' Gert Knepper says.

When no further explanation seems forthcoming, I asked who had placed the ad. 'Was it Theodor, or did Paula do it?'

'She did it.'

Otherwise, Gert knows only that the letter was sent to a post office box, and that the advert had been in a Catholic weekly. He doesn't have the actual text itself.

9

THE REPORTER WHO SETS FORTH with an open mind is bound to reap more than they bargain for. That axiom certainly applied to me, once I'd passed Eijsden, where *De Maas* becomes *La Meuse* once and for all and the river valley grows increasingly narrow.

The foretoken comes at first light, in the form of cooling towers. I am eating breakfast at my hotel in a virtually deserted Chateau Neuville. It is the day after Epiphany, and there is snow in the air. The window frames a view of the riverside occupied entirely by three concrete industrial colossi. I have to get up and move closer to the window to see that they are emitting a grey vapour, as though a witch's brew is bubbling under the lid. They're only blowing off steam, yet the sentinels of reactors Tihange I, II and III are there, radiating a vague brand of grimness.

A bit later, driving my rental car along the Meuse, I keep an eye on the towers in my rear-view mirror for another kilometre. Then they make way for an alum plant, a foundry, the zigzag of conduits around a chemicals complex. Asphalt and water run parallel here;

I pass inland waterway freighters on their starboard side. Meanwhile, on my port side, a row of smokestacks sails by. They poke up above the walls of rock, exhaling their foul breath in wisps of black and aqua regia yellow. I wasn't expecting it, but around every bend is yet another factory. Or a silo, a conveyor belt, a power pylon, a steel catwalk or a hoisting crane at a quayside. The Meuse Valley in the Walloon provinces, it turns out, is an industrial landscape, 100 per cent man-made.

How many signs and signals must come along before one switches track? For the next hour or so, I'm somehow able to ignore all the markers along the road. My only excuse: I have come here for Stone Age man in his natural state, a hairy nudist who produced no more indecomposable waste than the chips he knocked off his flint.

LIKE AT THE roulette table, reportage revolves around chance. Stumbling upon what you weren't looking for – serendipity – may produce unexpected riches. Sometimes they lie scattered on your path, sometimes they're right under your nose. But I am concentrating too hard on the sultry computer voice that directs me to my destination: the village of Engis, just outside the city of Liège.

In 1829, two fossilised skeletons were found: Engis I and Engis II. They lay in a complex of grottoes, amid the carcasses of cave bears and woolly rhinoceroses. My map calls them *Les Grottes Schmerling*, after their finder. But the government has closed the Schmerling caverns – even spelunkers no longer dare to enter them, due to the danger of cave-ins – and so I set course for the statue of Schmerling himself, behind the church of Saint-Etienne.

The fate of Philip Carel Schmerling, born in the Dutch town of Delft, fascinates me. More than Eugène Dubois, Schmerling might rightly deserve the title of 'founding father of paleoanthropology', were it not that he himself had absolutely no idea – could have had absolutely no idea – of the real meaning of his discovery. He was too many decades ahead of his time, and that was his great misfortune.

As an anatomical pathologist, Schmerling recognised that his fossils had to be extremely old. In 1833 he concluded that they must be of creatures alive before the Deluge: members of a separate, extinct branch of humankind.

Schmerling's explanation fell on rocky soil and did not take root. The mere idea of a primordial human type – a quarter of a century before Darwin's *On the Origin of Species* – sounded insane. According to the Old Testament, God had created man (note the singular) in His image. Not so much later – about four thousand years before Christ – he had opened the gates of heaven in order to wash away His creation, with the exception of Noah and his ark full of animals. Schmerling's view was diametrically opposed to this prevalent, albeit fanciful, picture of things, and so he was labelled a blasphemer. The doctor died in 1836 as a wild-eyed storyteller, a charlatan.

Nevertheless, his *Recherches sur les ossements fossiles découverts dans les cavernes de la Province de Liège* reads as matter-of-factly as a police report: 'The first cave at Engis lies some 60 metres above the Meuse. The entrance is triangular, with a base of 95 centimetres and a clearance of 80 centimetres.' Like a spider on its thread, the 39-year-old physician descends into a subterranean chamber: 'With the aid of a cord, attached to a tree, I lowered myself down.' There in the depths he came upon a femur and the spine of a human. The cave was dripping with moisture and provided access to other side-chambers. After hours of thorough investigation, by the light of a miner's lamp, Schmerling climbed back up to the daylight in the possession of two human skulls.

The remains found at Engis I are those of an adult, and at Engis II those of a two-year-old or three-year-old boy with an anomalous cranial shape that does not necessarily entail a deformation. During years of study, the physician became convinced that he had discovered an extinct, primitive brand of human being, a caveperson from the age of the cave bear.

One hundred and seven years later, the bones were subjected to a new autopsy. Professors at the University of Liège removed Engis II from its box and started their measurements. They arrived at the conclusion that the skull was that of a Neanderthal child. The year was 1936, exactly one century after the death of the bones' finder. Retroactively, Philip Carel Schmerling enters the footnotes of history as 'the first to discover an extinct hominid': Engis II appears to be the very first specimen of a Neanderthal ever found.

And it came from the Meuse Valley.

I FOUND IT striking to see how willy-nilly were the names given to newly discovered hominids. In some cases the location took precedence, in others some characteristic trait. *Homo heidelbergensis* refers to a sandpit close to Heidelberg, where in 1907 a jawbone was found with teeth so coarse that it was relegated to a separate hominid species. *Homo habilis* ('the handy human') was presented as such to the world in 1964, and was so named because the Leakey family saw that he was a fairly accomplished stonecutter.

Of all these names, 'Neanderthal' is the one most heavily laden with irony. The original suggestion of '*Homo primigenius*' (first-born human) was rejected, after all, in favour of a name that actually honoured a theologian: Joachim Neumann (1650–1680), who had

Hellenised his surname to 'Neander'. This devout composer of hymns
often went walking in a valley that was later named after him, the
'Neandertal', where, some two hundred years later, in 1856, the first
bones were excavated of what was officially seen as *the* Neanderthal.
The Belgians, however, promote Engis 11 as the holotype. Schmerling
found the boy's skull twenty-five years before the unearthing of the
adult male skull from Neander's beloved valley close to Düsseldorf.
If one wishes to do service to history, therefore, one would do better
to speak of the 'Maasdaler' – or the 'Meuse Man'.

That nationalist sentiment can play a role in something as trans-
border as the search for the origins of humankind seems not only
amusing, but amazing. Paleoanthropology scurries from one find to
the next. The discipline is dependent on chance hits. Like evolution
itself, hominid studies seem driven by the two-stroke motor of cap-
rice and coincidence.

But does the road it follows lead anywhere at all? A century and
a half of fossil hunting has produced the caricature of an ape travers-
ing five or six phases to become an uprightly mobile human. The skull
becomes increasingly rounder and larger, the physique more slender.
The series of silhouettes suggests a direction: after a slow start off the
blocks, we now stride forward with our chins held high. But what,
then, are we to do with the newly discovered hominids that don't fit
into this picture – like the undersized Flo, with her minuscule brain?

Concerning the ongoing tug-of-war between the acceptance and
rejection of new skull finds, our class has come across an article titled
'Receiving an Ancestor in the Family Tree'. The suggestion here of
intractability pretty much says it all. As a rule of thumb, paleoanthropologists assume that it takes a full generation for the dust to settle
around the discovery of a new hominid type. Lucy was the exception:

thanks to her appealing nickname, she was able, upon her discovery in 1974, to hitch a ride on the Beatles' fame. If she had operated under her catalogue name, AL.288-1, things would no doubt have been different.

At our behest, student Bob has read *The Fossil Chronicles*, a book about the forbidding reception given to archaic hominin types. Perhaps the most striking was the rejection and later embrace of the Taung Child. After its presentation in *Nature* on 7 February 1925, this little South African skull was dismissed for the next two decades as inconsequential. The failure to appreciate its importance ultimately caused the finder, Raymond Dart, to descend into depression in 1942; he stopped showing up for work at the University of the Witwatersrand in Johannesburg, and barely left his bed. But in 1945 the Taung Child was suddenly rehabilitated: in view of his combination of a flat (humanlike) face and the more pongid (apelike) back of the skull, he was retroactively accorded the status of missing link.

Raymond Dart arose from his lethargy, and in 1947 launched a daring new theory: the Stone Age, he said, was preceded by another crucial period: the Bone Age. More than two million years ago, according to Dart, the members of the Taung Child's clan learned to use the bones of their prey as clubs and spears. The worked ulna of a gazelle was, in Dart's view, indisputably a 'dagger'. It was not the transition from crawling to upright locomotion that marked our moment of becoming human, but the awakening of our bloodlust. In his Taung skull, Dart saw a foreshadowing of humankind's cruelty, unparalleled by any beast. He was 'a bone-breaking ape' who performed 'berserk' deeds to acquire and prepare his carnivorous diet.

The still-fresh memories of World War II gave wings to Dart's ideas. In unacademic prose he warned humankind against its own true nature, as inherited from the Taung Child. *Homo sapiens* bears 'the mark of Cain' on its forehead, he said; within the animal kingdom we are not simply murderers, we are fratricides.

In 1984, visitors to the *Ancestors* exhibition in New York proclaimed the Taung Child to be 'the world's most important fossil'.

Spectacular as the discovery of an early skull may be, eligibility for adoption apparently requires that it be found at a moment when the social tide is in its favour. If not, oblivion is its lot. See here the fate of Engis II and its finder, Philip Carel Schmerling, whose visionary insights were dashed to pieces against the dogmas of the church.

IT IS JANUARY 2017. I've had to farewell my students and now am following my own bearings. My destination, Engis, is no random stopover. The idea of following the river as a flowing element in the landscape came up during one of our last sessions. Someone coined the term 'reportage-fleuve' as the non-fiction counterpart of the 'roman-fleuve', or 'river novel', which describes a series of linked stories. The Meuse could serve as the cord on which to thread a string of loose beads, or chapters.

In practice, though, things have a way of going differently. For starters, the church of Saint-Etienne is not in Engis, but in neighbouring Awirs, and not down along the river, but behind an abandoned strip mine halfway up the hill.

The square beside the church is deserted. The bronze bust of Philip Carel Schmerling turns out to be hidden from view behind a wooden shed. The shed – open only at the front – houses a nativity stall. On a bed of straw behind an iron grille I see the Three Wise Men, Joseph and Mary, a cow and two sheep, all gathered around a wicker basket with a baby doll in it. There is no crib. To get to Schmerling's bust, I have to climb over the branches of the stripped municipal Christmas tree, which is already lying on its side.

An inscription reads: 'Founder of Human Paleontology'. There's no doubt about it, the University of Liège is behind this. Honouring Schmerling as the progenitor of paleoanthropology seems both clear and innocent, yet it is a gauntlet tossed at the feet of Leiden University, which claims more recent assets in the form of Dubois' fossils. In Leiden's view, Engis II is no more than a starter, not the pièce de résistance.

But why is that?

A bronze plaque is attached to the pedestal, with a quote from Schmerling from 1833: 'In the district of Liège, humans lived as contemporaries of the cave bear and other extinct species.'

And a handsome fellow he is. His gaze is calm, and the lines of his face speak of an easygoing disposition. Atop his head lies a skullcap of snow. An arrow thick as a man's wrist indicates the starting point for a 'Schmerling Route'. I am just about to head down that path when I hear voices behind me. Speaking at a hush. In French, close by. Turning, I see only the back of the nativity stall. I hurry to finish my notes ('snow on the shoulders as well, like epaulettes'), then walk around the side of the shed. But no one's there. The church square is deserted. Still, less than a minute ago, I heard two men talking.

When I take another look at the nativity scene, I see that Jesus is gone. Basket and all. For the rest, none of the wooden adorers have moved an inch – I'm sure of it. Hurriedly, as though I've lost something myself, I start scanning the square. Close to the dumpsters on the far side, there is no one. The cemetery behind Schmerling is deserted too. But then, across the street, I pick out two men in front of the local pharmacy. One of them is carrying something; he nods farewell to his companion and disappears down an alley. His friend heads straight for me – or, actually, straight for his Mercedes, which is parked along the street.

'Monsieur?' I ask.

He is, it turns out, the pharmacist, just about to begin his rounds delivering medicine to the homes of the ill. He also has the key to the nativity stall, and yes, his brother-in-law had just come by to pick up the basket with the Christ child; the rest will go back into storage this afternoon.

'The only thing I know about Schmerling is that he discovered our caverns,' he tells me. 'But the person who knows all about him is Madame Gérard. Why don't you ask her?'

EVOLUTION AND EVERYTHING that has proceeded from it to date bears the mark of the pointless, the aimless and above all the senseless. Darwinists claim that we as a species arose by accident, the result of an endless series of collisions. In the molecular soup of the cosmos, things just happen to gel, up to and including the formation of sentient beings. But those same beings are sorely mistaken if they think they see meaning or sense in the nature from which they originate – or, worse yet, if they think they see signs of a higher power. That renders them pathetic.

Still, that is exactly what people do all the time: they prefer to abide by the stories they themselves invented beforehand. Sacrifices are made at the spring, dances are done around the fire, prayers go up from the mountaintop. Damien Hirst overlays a skull with diamonds, and we all want to see it.

Which skull from which branch of the human family tree becomes a celebrity and which one doesn't depends only in part on its rarity, intactness or age. What counts most of all is a good story. In 1993, Leiden made the skull that Dubois found a worldwide topic by organising a huge centennial celebration: a celebration with international appeal, all for the Javanese ape-man. *Man-Ape, Ape-Man* was the title of the Dubois commemoration. During an exhibition in the seventeenth-century Pesthuis infirmary in Leiden, the ape-man's remains went on public display for the first time in the century since their discovery. His Royal Highness Prince Bernhard wrote an opening address, Richard Leakey compiled the catalogue. The ageing paleoanthropologist, an attraction in his own right, was flown in from Nairobi. Leakey spoke of a 'milestone in human history'. At the heart of the exhibition stood a pillar. On top of it, at eye level and beneath a dome of bulletproof glass, were the skullcap, the femur and the molar. At a respectful distance, from behind a railing, the tens of thousands of visitors were able to gape at the holotype of *Homo erectus*.

'My job here is to guard the *Night Watch* of anthropology,' a museum attendant told a journalist.

In *The Fossil Chronicles*, Bob had come across a term that seemed to us to fit the Dubois centenary promo to a tee: 'paleopolitics'.

THERE COULD BE NO greater contrast with the way Philip Carel Schmerling's memory is guarded. The inhabitants of Awirs place their local hero – at least between Christmas and the Twelfth Day – literally in the shadow of the Bible story.

But I'm in luck: the case of the missing Jesus from the nativity stall has given me just the push I need. Before I know quite what's happening, I'm standing at the hygienic white counter of Pharmacie Wera, talking to Madame Gérard on the phone. Then, while doing my best to mould a few French words into something like a sentence, improvising furiously, fortune smiles on me again.

'You don't happen to speak Dutch, do you?'

Until her wedding day, Madame Gérard says, her name was Wilhelmina van Loon. She hails from the town of Maaseik in Belgian Limburg, and yes, she is bilingual. 'Schmerling was trilingual,' she observes. 'He spoke German too.'

Half an hour later, we continue our conversation at her kitchen table. Wilhelmina van Loon lives up above the Meuse Valley: the road to her old farm climbs by way of a few curves to a plateau where a stiff breeze is blowing. 'The air up here is cleaner,' she says.

She gives me leave to take my shoes off in the hall and I enter the living room, where the sideboards are covered with carved wooden statuettes. Mrs Van Loon has dark-grey curls and rosy cheeks. Instead of a cardigan, she wears a fleece. She speaks of karst and calcite. Pleistocene and Holocene. And while she does, she checks whether I know about the hercynite fold that lifted this part of Belgium to above sea level – or don't they teach you that at Dutch schools anymore?

'It must have been complete chaos around here.' Sliding her hands towards each other like a pair of flatirons, she makes a few large billows in the tablecloth. 'See those mountain ranges? The Meuse had to find a way to push through those.'

'When was that?'

'Three hundred million years ago.'

On my way here, it seems, I had driven past coral reefs, petrified outcroppings from the Cretaceous sea. The river cuts its valley from the earth, and so the oldest formations have come to the surface here and there. Unquestioningly, Mrs Van Loon assumes that I, as a Dutchman, must know about the mosasaurus, a petrified reptile named after the Meuse – *Mosa* in Latin. Its more prosaic name – lizard of the Meuse – does no justice to its size: it was a gigantic predator almost twenty metres long, related to the largest of the still-existing monitor lizards. (*Komodo dragons!* I think, but I don't interrupt her.) The skull of this riverine monster was dug up in 1770 at Mount Saint Peter. A quarter of a century later, in 1794, a group of Napoleon's grenadiers captured '*le grand animal de Maëstricht*' for France, for which they were rewarded with six hundred bottles of wine.

I'm glad that I already knew about the mosasaurus: I had admired its impressive array of teeth once at the Muséum d'Histoire Naturelle in Paris. I was even more fascinated at the time by the Netherlands' persistent attempts to get '*le grand animal de Maëstricht*' back from the French.

'The mosasaurus lived in the Late Cretaceous period,' Wilhelmina van Loon says. 'About seventy million years ago. It was like the Bahamas here at the time.'

The image takes me aback for a moment, but I realise that I am meant to imagine palm trees, a subtropical climate. 'So you're a geologist?'

That, she thinks, is giving her too much credit. Yet Madame Gérard did study geology at Liège. 'Only as a non-matriculating student, though,' she adds. 'And I only started at an age when other people go into retirement.' Even as a child, she had collected stones – her pockets were always filled with them. 'But my father died young, so there was no money for me to continue my studies.' She depicts her adversities without complaint. By moving in with her husband in Engis – fifty years ago now – she crossed the language rift that divides Belgium like an earthquake fissure. Although she enjoys teaching, she's not allowed to do that in Wallonia: her teaching certificate is valid only to the limits of Flanders.

'When my daughter was fifteen, she said to me: "I don't want to hear you complaining later on that you gave up your whole life for us."'

By a well-timed stroke of luck, Madame Gérard had become the guardian of a grotto. One Sunday in 1989, her brother from Maaseik said he wanted to show her a cave further up the valley. It turned out to be barricaded. There was a fence in front of it, with a padlock. Asking around in the neighbouring village revealed that no one felt like watching over the key anymore. Madame Gérard volunteered, and before long she was not only the cave's keeper, but also an expert on Schmerling, an auditor of classes in geology and minerology at the University of Liège and a guide on fieldtrips for (Dutch-speaking) students.

'I enjoy providing information, as I'm sure you've noticed already.'

With regard to Schmerling, she tells me that his father was a religious refugee, a liberal Protestant from Vienna. That's why Philip Carel was born in Delft, on 2 March 1790. After Napoleon's defeat, he joined the army of King Willem I as a medical officer. He married in 1821, and afterwards moved with his family to what was then still the southern Netherlands, to Liège.

'They lived on Rue Hocheport, close to the station. That was not at all a fancy neighbourhood at the time.'

Schmerling felt called to alleviate the suffering of local miners who had contracted lung disease from inhaling dust. While paying a house call to a sick worker one time, he saw children playing with a few bones. 'They tossed them in the air to see how they would fall. Schmerling noticed that they were human bones, and asked where they came from. "We found them," one of the children said. "In the cave here, behind the house," said the other.' This 'cute story', Wilhelmina says, had led to the discovery of the first prehistoric human, and to Schmerling's denunciation.

She herself feels that people should not be kept in the dark; a priest doesn't have to know anything about fossils, but if he doesn't, he shouldn't stick his nose into it either. To offset all those believers who draw their knowledge of geology from the Book of Genesis, she dedicates herself to the memory of Dr Schmerling.

'Is that going to make any difference?'

'Not really.' She sounds resigned. There are other problems, more urgent ones. She supposes that I caught a whiff of what she means on my way here.

Now that she mentions it – yes. Suddenly I believe I can smell the dust that must have crept into my nostrils earlier in the morning. 'Coke fumes?' I ask.

'Mixed with other flue gases.'

Our conversation turns to concentrations of particulates that exceed the legal norm. There is no use in hanging your laundry out to dry in villages like Engis; clean sheets come back in dirty again. Then she tells me about her son: because he suffered from respiratory problems, the family had moved from the village down the hill up here to this farm. On government orders, the smokestacks along the Meuse were at one point built up dozens of metres higher, so that their long necks just barely stick out above the hilltops.

'These days, up here on the plateau, we look out over the black plumes of the refineries and the yellow fumes from the paint manufacturers. The only thing missing is red smoke, then we'd have the Belgian flag waving permanently over the fields.'

My hostess looks amused. Engis-sur-Meuse, she clarifies, is better known for its nineteenth-century zinc and lead furnaces than for

Schmerling's skulls. 'Hardworking Belgium', that's how the industri-
alisation of the Meuse Valley was referred to. Modernity and progress
had triumphed over nature – until cows and sheep started dying if a
windless stretch lasted more than a couple of days. And in 1930, people did too. She unfolded for me an account of the wave of fatalities
in Engis that year, during a particularly foggy period from 1 December
through to 5 December. The mist was permeated with ammonia and
sulphur compounds, a sourish, clammy fog that made people gasp
for air. I had heard about the Great Smog of 1952 that took the lives
of thousands of Londoners, but never about the deadly fog along the
Meuse. It turns out that the inhabitants of Engis, or in any case several
dozen of them, were the world's first official victims of air pollution.

   'There's a memorial statue for them in the village too,' Wilhelmina
van Loon tells me. 'It's a girl, looking down at the ground. Right beside
town hall.'

   She rises to her feet, rummages through a drawer in the built-in
cupboard and comes back with four packages of medicine. Each one
contains a blister pack of ten potassium iodide tablets. Written on each
is: 'Use Only in the Event of a Nuclear Accident / *En Cas d'Accident
Nucléaire.*' Local pharmacies handed out these pills free of charge
to all those living within a 50-kilometre radius of Tihange. Iodine
consumption is supposed to prevent thyroid cancer in the event of
radioactive fallout; Madame Gérard regards that claim with scepticism. There is a folder that goes along with the medicine, with practical
tips for Judgement Day, brightened up with pictograms:

- remain indoors (or go home immediately);
- close all windows and doors;
- tune in to the news on radio or TV.

In the event of an evacuation, Wilhelmina has been directed to take the fast road out of town, but she already knows that she will choose a different route. 'There are too many of us to all escape at the same time.' She puts away the brochure and the medicine packages. Once the table has been cleared, she says: 'I often think about the primitive humans. They lived from nature, and with nature. With only a deer antler and a couple of flints for tools. How did they do that? What went on in their minds?'

LOUISE, VINGT ANS is the title of the statuette commemorating the fog fatalities of 1930. Young Louise is kneeling on a polished chunk of granite, her hands in her hair. Her face averted.

> *In memory of the sixty, young and old, who died in Engis during the atmospheric mishap of December 1930.*

'Atmospheric mishap' makes it sound like a natural phenomenon. Such audacity. As though Louise at twenty would have choked in a normal fog of water droplets, no matter how thick it was. This angers me. Here, close to the steel girders of the Engis Bridge, the factories poke up like a riverine forest; the bare trunks of fifteen fuming smokestacks can be seen in a single viewfinder. And what is the motto on the plaque, beneath Louise shown at the moment of her death?

> *All human undertakings, including industrial ones, are prone to improvement.*

All right, I'm willing to go along with the view that humankind is the product of accident. But if that's it, then the accident is of a very specific kind: derailment. Clever as we are, we have succeeded in throwing the switches of evolution and running ourselves right off the rails of nature. As the self-appointed 'crowning glory of creation', we have succeeded in polluting the planet, including its atmosphere, so badly that it is killing us. Welcome to the Anthropocene.

10

'*FLORES MAN HAD REDUCED BRAIN*'
Tokyo – 17 April 2013.
The brain of the dwarf-like humans who lived some 18,000 years ago on the Indonesian island of Flores probably shrank in the course of evolution . . . The finding would seem to indicate that Flores Man was a dwarf version of the archaic human species Homo erectus. This conclusion was reached by researchers from the National Museum of Nature and Science in Tokyo. (NU.nl online newspaper)

WHILE THE AVERAGE reader skimmed right over news items like this, the interested parties to this 'brain debate' were at daggers drawn. Ten years after her disinterment, Flo had grown to become the focal point of a doctrinal clash. Was evolution actually *going* anywhere?

'Yes,' one camp said. In the course of hundreds of millions of years, life on earth has developed from single-celled animals to mammals; evolution tends to produce increasingly complex lifeforms.

'No,' their opponents said. Species of fish living in underwater caverns lose their eyesight within only a few generations. Certain functions may die off quite quickly, and the adaptations can go in any direction.

But now, in the case of LB1, the discussion had suddenly turned to hominids. 'We' had apparently risen up from the animal kingdom in a few firm strides, and gradually accumulated a much greater brain capacity than those we had left behind. Or was this only a random indicator? Might *Homo sapiens*, in principle, just as readily become weaker, smaller and stupider?

As master's candidates in philosophy, Bob and Lian have done some thinking about the term 'devolution', a natural phenomenon that doesn't coincide with evolution-in-the-opposite-direction. As with a ratchet, Bob explains, the possibility of turning back was blocked. There was no rewind button. This was not to say, however, that after a period of growth organisms could not shrink again. As to whether this also applied to becoming more intelligent and then more obtuse, the jury was still out.

In Flo's case, the experts were talking about a 50 per cent reduction in brain volume. If LB1 was a direct descendant of Dubois' Java Man (*Homo erectus*), then her brain must have shrunk on Flores from a meagre 1000 cubic centimetres to a little over 400 cubic centimetres.

'It's possible,' the Japanese researchers stated in April 2013. The result of their study would have been a huge boon to Mike Morwood, had he himself not already rejected the scenario of evolutionary dwarfism. 'It can't be,' he realised. *Homo floresiensis* had followed an evolutionary path of its own, alongside *Homo erectus* and *Homo sapiens*. She was a fresh twig on an ancient branch, which Morwood suspected had stretched itself out over a period of more than two million years, and from Africa to Asia.

LESS AND LESS remained of the romantic view with which we first approached paleoanthropology. Those who looked for skull-hard evidence to bolster philosophical notions had indeed been serving a higher cause. But meanwhile, the entire field was under attack. The 'old school' stuck to its attempts to demonstrate humans' unicity by finding out at which moment and in what way the genus *Homo* had split off from that of the apes, but the younger generation felt that was nonsense. 'There is no difference between humans and animals,' was their rebuttal. 'So stop looking for it!'

With her view that 'fish are every bit as competent as humans', José Joordens did not stand alone. One of us found a quote from the editor-in-chief of *Nature*: 'Giraffes and dung beetles are just as good at what they do as we humans are at what we do. Period.' The whole idea of a missing link was based on a misconception: there was no yawning chasm between human and animal, and so there was nothing to be bridged. In a reinterpretation of the importance of Java Man, Eugène Dubois was now presented as 'the first person to present proof that we are not above nature, but a part of it'.

Emphasising the singularity of *Homo sapiens* was passé, out of fashion. It was said to be the product of unwarranted arrogance, a detrimental form of self-conceit. Of anthropocentrism. As early as 1999, Paul Sondaar had given voice to this rising sentiment when he said that he opposed 'the established bias: the endless pointing out of what makes man so different'. Sondaar got what he asked for. The ecologically correct position these days amounted to toeing the party line: endlessly pointing out that *Homo sapiens* is an animal like any other. The bonobo is said to be almost our equal, seeing that 98 per cent of its DNA is identical to ours. In this case, my own curiosity was immediately fixed on that remaining 2 per cent. Unlike society at large; the point was rapidly approaching when an ape who made a selfie could acquire the copyright to that image. 'Humans have to relinquish our power over the animal kingdom,' a candidate MP for the Dutch Party for the Animals proclaimed.

A philosopher of science from the University of Leiden felt that, in the year 2018, it was 'hard to persist in the claim that humans with our mental powers are fundamentally different from other creatures on this earth'. In defence of this point of view, a whole procession of animals is usually trotted out, to demonstrate one by one that they can do something that we humans thought made us unique. A rat expressing 'regret', a capuchin monkey acting 'jealous', a chimpanzee 'thinking ahead', or a crow 'coming up with' ingenious solutions to the problems with which it is presented. Phenomenally bright and moving. Still, this menagerie of animal exhibitions didn't convince me. As though it made no difference who was observing and who was being watched. Even without the fables, it is not hard to see that we use animals as living projection screens for our own troubles and cares. Who posted videos like that on YouTube, anyway – the animals themselves? To hold up a mirror to us?

The fact that humans are animals does not mean that the converse – animals are human – is also true.

In his report on the collective puzzling over the plaster casts of skulls during our visit with John and José, Bob remarks: 'The idea that people are more than animals seems, in their field, to be a thing of the past. But don't their own controversies and feuds illustrate what it is to be human?'

PALEOANTHROPOLOGY MAY HAVE BEEN relegated to the function of doormat these days, but it was certainly not devoid of liveliness and tumult. The Bone Wars of the nineteenth century, in which American dinosaur experts combatted each other with bribery, vandalism and forgery, were living on, having barely abated, when it came to Flo's skeleton. For the rest of his life, like Eugène Dubois before him, Mike Morwood suffered under his triumph – as though a sort of curse of Tutankhamen rested on those who exhumed archaic hominid remains.

Morwood, the son of a baker from Auckland, New Zealand, was a fervent collector of samurai swords. His critics never missed an opportunity to remind him that his training was in the field of archaeology: why, then, didn't he stick to studying hand axes and hearths? By venturing into the territory of craniology, Morwood had burst the banks of his own discipline. Aboriginal rock drawings from Australia's northern coast were his specialty. Looking out over the sea there, though, he had begun asking himself where Australia's first inhabitants had come from. Indonesia? But how? Drifting from island to island aboard rafts, all the way past the Wallace Line? And was Australia simply the next stop among many?

In Classroom 0.04, we had put together a sizeable dossier on Morwood, including photos. Judging from his appearance, you would suspect him more of nonchalance than of bad faith. Mike was a prematurely greying fellow with longish hair. Upon getting up in the morning, he first greeted the sun, as prelude to his aikido exercises. During his fieldwork, he wore a cowboy hat.

As soon as Indonesia gave him an excavation permit, in 2003, he'd had the pits in Liang Bua deepened and shored up with bamboo struts. His workers found the skeletons of stegodons and a circle of stones split by the heat of many fires – probably a hearth. In his mind's eye, Morwood saw troglodytes chewing on the meat of pygmy elephants. But among the countless fossils – of Komodo dragons, bats and giant rats – there were no remains of human cave dwellers. Up until the very last day (some sources say the next-to-last day) of the excavation season.

The story behind the great breakthrough goes as follows: on the morning of 6 September 2003, Benjamin Taurus climbed down to where he was working at the bottom of a six-metre-deep pit. Squatting, a trowel in one hand and a brush in the other, at 9.40 a.m. he scraped aside a layer of earth. The sweep of the brush revealed the brittle bones of a unique, as yet unknown hominid. The skull was as soft and moist as papier-mâché. Benjamin Taurus made way for the excavation foreman, Thomas Sutnika, who first reinforced the remains with nail polish. The left brow ridge was damaged, and it took the rest of the week to extricate the skull in its entirety. What they had discovered was not just a separate head, but an almost completely intact skeleton.

The skeleton brought up rib by rib was so small that Thomas Sutnika was convinced they'd found a toddler.

All the credit went to Mike Morwood, even though he was in Jakarta at the time of the find. His description of *Homo floresiensis* in *Nature*, one year later, sent shockwaves through the international community. Nothing comparable, he said, had ever been found on the planet before. The find cast a whole new light on the question of what it meant to be human.

In the days immediately following the revelation, an ever-swelling wave of rejoicing rolled around the world. An Australian philosopher predicted that Flores Man could change humankind's ideas about itself in the same way as did Copernicus' insight in 1530 that the earth is not the central point of the universe. Two venerable paleoanthropologists were also generous with their praise. One spoke of 'the most important discovery concerning our species in my lifetime'; another referenced *Star Wars*: 'These miniature people are not something dreamt up by George Lucas. They actually existed.'

Morwood let the superlatives roll over him. He realised that LBI –
as a woman of about thirty years – was hardly larger than his own
three-year-old daughter, Julia. Especially for her, he has a replica of
Flo's skull made; it was about the size of a tennis ball.

Soon enough, though, his bravura gained the upper hand over his
modesty. Morwood was restless, almost to the point of turbulence.
In interviews he insisted on referring to Flores Man as 'the hobbit'.
Why the nickname? Was he playing to the crowd? Yes, Morwood
explained. He'd had a hard time convincing his team to go along with
him, but the allusion to Tolkien's creation would keep them from
falling into obscurity.

MORWOOD'S FIRST CHALLENGER was Teuku Jacob. As the
grand old man of Indonesian paleoanthropology, he brought more
than half a century's experience to bear. Concerning Jacob, the story
goes that in the mid-1960s, while a university student in Utrecht, he
spent a few days in hospital for an appendectomy. When his PhD
supervisor, Professor Von Koenigswald, came to visit, Jacob wangled
from him the promise that he would give his Sangiran skulls back
to Indonesia. When Von Koenigswald failed to keep that promise,
his former student personally saw to it that he was no longer able to
enter Java.

Later, as a member of parliament, Teuku Jacob was instrumental
in drafting the *Cultural Property Protection Act*. On Jakarta's behalf,
he filed a claim against the Dutch government, demanding it return
the skull of Dubois' Java Man. 'Indonesia Wants Ape-Man's Bones
Back' was the headline in the *Leeuwarder Courant* on 16 September
1977. The next year, he and Von Koenigswald patched up their quarrel

with the handing over of several Javanese skulls (but not that of Von Koenigswald's beloved Sangiran IV, nor anything of Dubois').

In 2004, confronted with the tumult surrounding Flores Man, Jacob felt offended. The white Australian scientists, and Mike Morwood in particular, were guilty of orientalism: with their Western way of looking at things and their gigantic physical stature, they thought they saw something extraordinary in the tiny size of his compatriots on Flores. My oh my, there used to be a separate human species living here, a population of dwarfs with the brains of a chimpanzee. Hobbits!

Professor Jacob had Morwood declared persona non grata: he was a 'science terrorist', tainted by 'Western arrogance'. Jacob quickly confiscated the skeleton of LB1. The bones were brought to Jogjakarta, a good four hundred kilometres from Jakarta, where he locked them in the safe at his own laboratory.

Whatever else you might say about Jacob, he was consistent. Whether it was the skulls found by his mentor, Von Koenigswald, or Dubois' ape-man, he never missed a chance to lay claim to Javanese property – in exactly the same way the Netherlands wants to get 'le grand animal de Maëstricht' back from France.

But this time his attacks were aimed at the person of Morwood. And he received unexpected support from Jean-Jacques Hublin, of Leipzig's Max Planck Institute, who stated that Morwood's attitude did indeed smack of 'pure scientific neocolonialism'. Shortly afterwards, it leaked out that Professor Hublin, apparently in return for his display of support, was allowed to export two original LB1 bones for DNA analysis in Germany. Even though Hublin didn't succeed in cracking their genetic code, the manoeuvre effectively put Morwood on the sidelines.

Once Jacob had abducted Flo, which amounted to the kidnapping of a hominid (a first in the history of humans), he extended an invitation to Morwood's enemies to come to Jogjakarta. In February 2005, four of them gathered to examine LB1's remains. Afterwards, they stood on the steps in front of Jacob's 'Laboratorium Paleo-Antropologi' for a group photo, as 'the Pathology Group'. To the press, this delegation of sceptics presented a united front in support of Jacob's earlier statement that Flo was simply a *Homo sapiens* deformed by some illness.

In April of that year, Jacob set up an expedition of his own to the remote kampongs of Flores. Just like in the good old days, he set about measuring the chins, ears and noses of the local farmers. One by one, he lined them up for a photo against the background of a tape measure. 'Individual 26', from the untraceable village of Rampasasa, was the figure he'd been looking for: the man's receding chin and low forehead bore such a strong resemblance to the form of LB1's skull that he alone could have been living proof that the 'hobbits' should be relegated to the world of fiction.

The abduction of the bones from Liang Bua lasted a total of three months. Once Teuku Jacob finally released the skeleton, she turned out to have been manhandled: LB1's jawbone bore two deep incisions, probably from a knife slipping during the removal of the mould. In addition, the pelvis was broken.

'I know nothing about it,' Jacob said in a telephone interview with *USA Today*. 'At least, it didn't happen while the remains were in my care.'

On 17 October 2007, Professor Jacob died of liver failure and was buried 'with military honours' in Jogjakarta. In that same year, the barriers were taken down around Liang Bua.

*   *   *

WHENCE THE FERVOUR of Teuku Jacob and his cohort? Could offended honour, national or personal, be enough to explain the ferocity of this war of bones? We wondered whether there was perhaps more to it than patriotism, machismo or a combination of both. If you took a few steps back, like the painter before a canvas, then suddenly you were struck by what a remarkable activity this skull-reading was: humans tackling humanity.

One of the students draws the comparison with a fish being asked to describe water. Another sees a dog curling up in order to examine its own tail. The troublesome thing about (paleo)anthropology is that its practitioners inevitably coincide with their subject matter. In addition, their findings have to do with each and every one of us. Pure and simple. Lay a finger on the skull of an archaic human, and you have to answer to all humanity. The emotions surrounding LB1 had flared so bright that there had to be more to it than the reputations of a handful of hominid hunters. They may have been fighting for their personal honour and reputation, but the real stakes were many times greater than that.

AFTER JACOB'S DEATH, Mike Morwood was allowed back into Indonesia. At the suggestion of *National Geographic*, he accepted a physical challenge: along with a group of allies, he would ascertain whether hominids like himself could possibly cross the strait between Sumbawa and Flores aboard a bamboo raft. The journey was to be a fight to the finish, featuring men with sunburned necks and raw hands, all documented by camera crews in following boats. And, of course, they reached their destination.

In 2008, like a stubborn gadfly, the Pathology Group struck again: the members bundled their grievances in an indictment entitled

The Hobbit Trap: How New Species Are Invented. The gist of it was: under pressure from sponsors and driven by bald-faced ambition, Morwood had succeeded in presenting to the world the body of a handicapped woman as the holotype of a hitherto-unknown hominid. In fact he turned out to be an amateur who was 'unable to recognise the village idiot when she (albeit in fossil guise) stands before him'.

Just as Dubois, three-quarters of a century earlier, had accused his rival Von Koenigswald of being a forger, the Pathology Group accused Morwood of fraud and manipulation: a lower left molar from the jaw of LB1, they claimed, 'shows signs of recent dentistry'.

THE PATHOLOGY GROUP received a striking amount of support from Africa; their most vocal supporter there was Raymond Dart's most prominent student. Less than ten years earlier, in 1999, the Cradle of Humankind – an arid stretch of land with caves and quarries to the north-west of Johannesburg – had been placed on the list of world heritage sites. With United Nations support, a spectacular skull museum was built there on a hilltop, in the form of a beehive – the implication was that 'we' had swarmed out from there all over the globe. But with the discovery of Flo, the media spotlights were now back on Asia, and away from Africa. To the paleoanthropologists who since World War II had built their careers on African fossils, this was a slap in the face.

It was only a matter of time, therefore, before Richard Leakey entered the field of battle. His reaction to Flores Man was an indifferent shrug: 'We're talking about only a few individuals here. They certainly don't deserve a place in the story of our evolution.'

Richard Leakey was Morwood's most formidable opponent. This white Kenyan drew his authority largely from his parents. We draw up a family tree of the Leakeys – who formed a dynasty of their own within paleoanthropology – simply to keep from getting the various members mixed up.

Richard's grandfather, Harry, was a missionary in Kenya; his father, Louis, an apostate who had traded in religion for science. Louis Leakey became world-famous after marrying his second wife, Mary, a student of his who set up her own camp in Tanzania's Olduvai Gorge and found footprints of the first primate to walk upright. Their eldest son, Jonathan, took over the reins, followed later by his impetuous brother, Richard, who also brought his wife, Meave, and their daughter, Louise, into the family business. Their greatest feat was the discovery, in 1984, of the 1.6-million-year-old 'Turkana boy'.

In 2014, Angelina Jolie announced that she would be making a feature film about the life of Richard Leakey, starring Brad Pitt. The film never happened.

'All people are Africans' – that is the godless gospel according to Richard. When asked why the Creation account enjoyed so much more adherence on his beloved continent, he was wont to reply: 'Because we have neglected to preach evolution each Sunday.'

For Richard Leakey and his family, Africa is the incontrovertible cradle of humankind. If paleoanthropology has its own castle keep, then it rises up in the east of that continent, roughly around Mount Kilimanjaro. Seen from there, Flores is indeed in the back of beyond.

Richard Leakey on *Homo floresiensis*: 'You could almost say: so what? What does a small population of mini-hominids really say about us anyway?'

*   *   *

MORWOOD WAS DEJECTED. To get revenge, he needed a second round. He had to present new, supporting evidence: new teeth, bones or entire skulls of other 'hobbits'.

As soon as he succeeded in arranging a new excavation permit, he went back to Flores. But not to Liang Bua. He and his regular team lit down on the same hill where Father Verhoeven had once found 800,000-year-old stone tools amid the remains of elephants (both large and small, including a tusk that measured 2.8 metres). Morwood had the rocky slope chopped and raked away, in search of hominid bones or teeth. If he could only produce the one who made those tools – a sturdy *Homo erectus* of 1.8 metres who knew how to kill huge elephants? – then he would have Flores Man's ancestor.

Morwood had been a teetotaller for most of his life, but now he began drinking beer in the evenings. He would smoke a pack of clove cigarettes along with it. Meanwhile, his worries piled up. Particularly about his daughter, Jarla, who had been diagnosed with a brain tumour. She died at the age of ten.

Three years later, in July 2013, Morwood himself succumbed to cancer: he died at the age of sixty-two in the Australian town of Darwin.

11

THERE IS A Neanderthal deep inside me. Between 1 and 4 per cent of my DNA is composed of the surviving remains of *Homo neanderthalensis*, who became extinct 40,000 years ago. That is a hard fact that neither philosophy nor literature can iron out.

To get to know my own exact 'Neander level', I have to send a cheek swab to America. The Mormons in Salt Lake City would be all too delighted to have it, but that's ruled out. An alternative supplier of DNA test kits goes by the name of 23andme – '23' being a reference to the number of chromosomes we carry in every cell of our body. Their laboratory in Mountain View, California, promises answers to the questions that beset me. 'Our service helps you better understand who you are and where you come from.' Am I willing to pay $99 for that?

The digital order form – 'Bring Your Ancestry to Life' – states that we, as humans, are living in the 'genomic age'. The mission statement: 'We at 23andme want to help others to take a determined, well-considered step towards knowing themselves.' It sounds Nietzschean:

become who you are. Yet it is bigger than that: 23andme enters your unique 'me' into a database with more than a million other unique 'mes', shedding a new and increasingly bright light on the average person and their roots. After all: 'The story of human history is captured in our DNA.' This story of origins can be told using only four letters: A, C, G and T – one vowel and three consonants. They symbolise the four bases that form the steps in the winding staircase structure of the DNA molecule. In pairs, and in a long strand containing some 3 billion units.

Even though the Neanderthals vanished from the face of the earth some 40,000 years ago, their DNA lives on in us.

MODERN HUMANS	NEANDERTHALS
Flat faces	Larger noses
Slim, long limbs	Broad, robust posture
Advanced skills and artistic ability	Large eye sockets, presumably better vision

23ANDME CAN TELL me what percentage of my genetic make-up comes from the Neanderthal, and the extent to which I depart from the average. There is a caveat beforehand, though, specifically for people with African ancestry: if your roots lie in Africa, you may not bear a single Neanderthal gene.

More disclaimers appear under 'general terms and conditions'. Genetic information, I would do well to realise, is always 'permanent knowledge'. It is irreversible. There is no denying it, not ever again. An unwelcome result has the power to disrupt lives and cause worldviews to collapse. Were I to take the test along with my daughter, for example, it could turn out that I am not her biological father.

I leave the 23andme website and go to that of *National Geographic*. This one looks less loud, more scientific. While ordering the *National Geographic* DNA kit – for $149 – I decide to take part in their worldwide genographic project, 'GENO 2.0'. I miss out on the annual Father's Day sale, though, as Father's Day is still months away.

WHAT HAD FAILED with Flores Man – the extraction of DNA from residual bone marrow – succeeded with the Neanderthal. When there are no hairs or nails to be had, the roots of the molars constitute the most likely source of DNA; little survives of that, however, in the perishable climate of the tropics. 'Young' as LB1 might have been, her genetic material had wasted away.

To preserve DNA strands for tens of thousands of years, cold, damp caverns are needed, filled with mud and sludge. Stalactite caves in moderate climes, located along a river that deposits sediment, like *Les Grottes Schmerling*. An attempt to go back and pick up a DNA 'signal' from the skull of Engis II, however, failed. But in 1993, in a cave further up in the valley, the 'Child of Sclayn' was found – or in any case the lower jaw of an eight-year-old Neanderthal girl who lived 90,000 years ago along the banks of the Meuse.

She still had her milk teeth, but was in the process of shedding them. The root of one of her molars (SCLA 4A 13) was sawn off, pulverised and decalcified in a sterile laboratory in 2006. This resulted in 200 micrograms of marrow, from which geneticists at Leipzig's Max Planck Institute succeeded in plucking various strands of DNA. The Child of Sclayn had contributed to something few considered possible: the unravelling of the Neanderthal's complete genome, an achievement that thoroughly rototilled the skull-readers' field of study.

For the first time, a window was opened that allowed a comparison of the genetic structure of *Homo neanderthalensis* with that of *Homo sapiens*. What were the differences? And where was the overlap?

SCLAYN IS A village on the Meuse, up past the industrial filth of Liège, Engis and Tihange I, II and III. The narrow road perpendicular to the river runs uphill through forest to the cave of Scladina. This part of the Ardennes is famous for its enchanting grottoes, where families in boats are ferried along subterranean rivers, but the gate at the entrance – 'Grotte Scladina' – has nothing touristic about it. Certainly not in January. The handful of parking spots is reserved for researchers from the University of Liège.

I have an appointment with one of them, Dominique Bonjean, leader of the excavations since 1991. On the basis of a quote I had found from him, he had little choice but to receive me: 'Science is like life. If it isn't passed along, there's no point to it.'

Dominique Bonjean wears a faded red hoodie with white drawstrings at the neck. The stubble on his chin is the same grey as his close-cropped hair. However informal his appearance, though, he does not wish to be addressed by his first name.

Monsieur Bonjean invites me into his office. Upon entering the hallway, I see a display case on the left with a replica of the jaw of the *'Enfant Neanderthalien'*. We turn right into the workshop (the 'atelier', he calls it). Before visiting the cavern itself, he directs me to a chair at the table in the middle of the room. He himself remains standing, leaning against the counter.

'I've got till a quarter to five, then I have to pick up my son from school,' he says.

Bonjean has four children, but he doesn't appreciate my asking about that. As soon as I assume a meeker role, he launches into an animated monologue about the cave of Scladina as an advance out-post for the University of Liège. This is the only place in Belgium where excavation work is ongoing, and has been for some forty years.

'What is it we do? We lift the veil of prehistory and make it less mysterious.'

Jacket draped over his shoulders and keyring in hand, he leads me outside. We take a wooded path that climbs across the slope; the steps hacked into the muddy forest floor are slippery as ice. 'Caves,' Bonjean says, 'are the womb of mankind.' Without turning to look at me, he adds that Native Americans see it in exactly the same way: life springs forth from caves. 'Schmerling explored sixty of them. He knew where he had to look.'

We arrive at the foot of a metal fire escape, built from galva-nised gratings. The steps lead to a scaffold placed against a windowless facade, with a door in the middle. This, it turns out, is the entrance to the cave. Before we go in, Bonjean turns to me. I should know that Sclayn is not the only place in Belgium where Neanderthal remains have been found, he says. After Engis II, another important discovery was made in the caverns at Spy, beside a rivulet along the upper Meuse.

I tell him that I've heard about 'Spyrou', the main attraction at Spy: a wax caveman with a beard and green eyes, said to be one of the most realistic reconstructions.

Bonjean ignores my comments. 'The discovery at Spy dates from 1886, the year in which the Americans created Coca-Cola.' By not asking what he means by that, I give him the opening he is looking for. Bonjean imitates a workman emptying a bucket of rubble in the corner of a marl pit. And behold, suddenly he sees the skull of a hominid. That's how the first Neanderthal at Spy was found. 'But alas, the find was worthless!' Bonjean provides the running commentary for his own charade. 'The workman went back into the cave and there he saw another skull, sticking halfway out of the limestone wall.' He rubs his hands and comes to the point: 'You always have to know exactly where the find was made. Without a context, there's no story.'

The vestibule of the Scladina cavern looks like a factory control room. The grillwork that serves as steps outside extends into the cave itself, forming a floor that narrows to a footbridge to the back of the cavern. Pushing his glasses up onto the bridge of his nose, Bonjean opens the meter box and throws a switch. Construction lamps plop on along the railing. I see that I'm in a tunnel that dead-ends less than fifty metres away.

'The spelunkers who discovered this cave in 1971 could only get in lying flat on their stomachs, that's how high the sludge had piled up.'

Seeing them in my mind's eye, wriggling along like salamanders under a rock, I feel slightly queasy. Thankfully, Scladina cave has been excavated to well over head height in the meantime.

'That is precisely the work we do,' Bonjean says.

Ever since 1978, university students from Liège have been coming here to do fieldwork. In fact, Scladina is not so much being excavated as it is being spooned out. In the course of four decades, 10 per cent of the material has been examined. Centimetre by centimetre, with trowels and brushes. The deposits of sludge are brushed away, down to the solid rock. Around these mole holes, a winding tunnel has been drilled out. Lamplight accentuates stalactites hanging from the ceiling like icicles.

I'm allowed to climb down from the footbridge to one of the digging sites. There is a pair of paver's kneepads on the floor. Roughly at eye level, an ivory-coloured object is sticking out of the earthen wall. 'What's that?' I ask.

'From the looks of it, a cave-bear molar,' says Bonjean, who has remained on the footbridge. He aims his flashlight at it.

A cave-bear molar – I can barely believe it. I have the urge to pull it right out of the wall.

'It will come out by itself, next season,' Bonjean says. This exaggerated caution of his is starting to rankle me. What's with this nonsense?

As though reading my mind, Bonjean starts in about the paradox of his profession. 'By digging, we destroy the thing we're studying. So you have to do it right the very first time, because you won't get a second chance.'

A few steps further into the cave, between metre-markers 28 and 29, he uses his flashlight to point out a stubby stalagmite. Right beside it, in the circle of light, is where they found the Child of Sclayn.

Bonjean leans on the railing with both forearms, like a bridge-man watching a passing ship. He tells me about one of the students, Claire, who came up from the pit with a section of jawbone and two molars. 'It was a Friday, the sixteenth of July 1993. I was on supervisor's duty.' Claire was allowed to finish excavating her discovery on her own, 'something the archaeologists in Leiden would never have countenanced'. At the end of the day, when Bonjean took the fossil to his workshop, he still had no idea what kind of mammal it might come from. He called a colleague in Liège, who came to take a look that Monday. '*Merde*,' the man shouted. 'It's a hominid!' The countercheck, carried out in France, confirmed it: this was the jaw of an eight-year-old Neanderthal child.

Bonjean is about to show me something else, but then there is a crackling sound, followed by a loud pop. The lights go out – the fuses have blown. 'This damn humidity,' he says. Moving largely by feel, we make our way to the exit. 'At least the Neanderthals had fire. A lot handier than electric light, and you can warm yourself at it too.'

THE PACKAGE THE postman brings contains a black box. 'GENO 2.0' was printed on the lid, and beneath that, in little white letters: 'Your story. Our story. The story of humankind.'

There's a code glued to the inside of the lid. That code is me. It's the key to my 'deep ancestry'. The accompanying folder offers me thanks and congratulations. I am about to take part, with my own saliva, in a 'historic quest for all our origins'. Although I'm supposed to be staying close to home for the time being, it feels like I've signed up for an expedition. The brochure talks about 'the greatest journey ever': how our ancestors left Africa and ultimately colonised the globe. Together – due, apparently, to my participation – we will map the course of that exodus.

'How many different treks were there out of Africa?'

'What role was played by the Silk Road, with its caravans and bazaars, in establishing new lineages all over Eurasia?'

In the small print, I read that *National Geographic* has entered into a cooperative venture with the 21st Century Fox movie studio. That clarifies the emphasis on the word 'story' – the text is rife with it. *Story* and *history* overlap. That fascinates me; they become synonyms.

'The greatest history book ever written lies hidden in our DNA.'

'You will add a chapter to the story of humankind.'

No one is forcing me into this. I can go back. Still, I feel no hesitation; I have, after all, urged my students to put their all into their reportages. When you're truly invested, I've been telling them, sooner or later you, as a reporter, become tangled up in your own story. Staying out of range is not an option.

The instruction leaflet calls for me to scrape two hygienically packaged spatulas along the inside of my cheek. Maybe I was too rough with it: there is blood on one of them. That's all right – there's DNA in that too. The slide attached to the handle releases the heads of the spatulas, so that I can deposit them in two fluid-filled ampoules. Put the lid back on and mail immediately – to 21st Century Fox.

IN THE 'ATELIER' outside the Scladina cave, I put the thumb-screws on Dominique Bonjean. From a cupboard, he had removed a Tupperware container marked 'Object 405'. A plastic replica of the fragment of the Child of Sclayn's jaw – but I am not meant to handle it myself. Bonjean pulls up a chair. I assure him that I am deeply impressed that molar 4A 13 had supplied the oldest hominid DNA ever. But doesn't that render his work redundant?

'What do you mean?'

I tell him I fear that old-fashioned cave spooning will soon become a thing of the past. Paleo-geneticists with their four-letter formulas, it seems to me, now have a truth serum at their disposal. A few drops of it is enough to negate the guesswork of the old guild masters.

'Or to confirm it.'

Yes, I nod, that too. Still, he knows as well as I do that DNA technology is blowing one hole after the other in the accepted knowledge about primitive humans. The comparison of A, C, G and T sequences shows that the Neanderthal was not exterminated point-blank by *Homo sapiens*. This standard assumption of genocide had been current for more than a century. Wherever *Homo sapiens* shows up, the mass murdering begins. In one commentary on the disappearance of Flores Man, the idea came back: 'We exterminate. That's the way we are. Destruction is in our nature.'

But the DNA comparison tells a very different story: we did it with Neanderthals, and they did it with us. Whether or not the mating took place amicably, the fact is that both species impregnated each other, and their descendants were not sterile. Bonjean and I both have Neanderthal DNA in us.

'So what you're asking me is whether a revolution has erupted in our field? But you already know the answer. So why ask?'

DNA is merely a new tool, not a panacea. Bonjean brings up the subject of Oasis 1. The DNA profile of this individual, who died 38,000 years ago in a cave along the Danube in Romania, showed that he was the great-great-grandson of a Neanderthal and a *Homo sapiens*. Nice to know, but whether Oasis 1 hunted with a spear or with a club, no lab worker in Leipzig could say.

My host stands up, picks up a water bottle, a roll of paper towels and a screwdriver, and puts them on the table in front of me. The bottle represents the archaeologist, the paper towel the anthropologist, the screwdriver the geneticist. 'If there are no stone tools,' Bonjean says, holding his hands palm up for me to see, 'then you don't need an archaeologist' – and he snatches away the bottle. 'If you've got no bones, then the anthropologist can stay at home' – there goes the paper towels. 'And without a cold, wet cave where DNA can be preserved, you can say goodbye to the geneticist.'

He remains standing with the screwdriver in his hand. Archaeologists like him have skills no geneticist in a lab coat can simply shrug off. He can deduce the behaviour of Stone Age inhabitants. 'If

I look at a hand axe, I can see whether it was made by a Neanderthal or a modern human.' *Homo sapiens* made elongated axes and scrapers; *Homo neanderthalensis* more compact ones, but sharpened on both sides. 'Undoubtedly, they learned from each other. Copied the tricks. That has nothing to do with DNA.'

But the comparison of their respective genomes has revealed their sexual contact, I object. When one talks about behaviour, that seems to me more intimate – and also more revealing – than their stone-cutting technique.

Bonjean feels that the cultural expressions are more important. The cave dwellers at Sclayn ate hares and wild goat – he has proof of that. They had their own quarries and working areas, and made use of resin and ochre. His greatest find has to do with the discovery of little black globules in the Scladina cavern. The first one he picked up he had accidentally squeezed between thumb and forefinger – a black powder was all that remained. 'I have a total of fifty-two of those pellets.' They were made of manganese oxide.

In the Ardennes, that black stone was only found on the surface at a spot forty kilometres from Sclayn. There was no way the water of the Meuse could have washed it into the cave. 'So the manganese was brought here. But by whom?'

Monsieur Bonjean leans forward and brings his face close to mine. 'The Neanderthal. And I also found out what he used those pellets for.' Here, as though in slow motion, he draws the fingertips of both hands across his cheeks. 'For make-up!'

Maybe they used it as camouflage while hunting, maybe as part of their rites of war. 'Why did they do that? We don't know. But the Neanderthal painted themselves.'

*   *   *

I WAS STUMPED. The deeper I delved into the work of the hominid seekers, the more the image of the echoing well came to the fore. What the diggers heard – wasn't that the echo of the very same story they'd used to interpret their own discoveries?

The arrival of DNA as lie detector didn't keep the paleologists from wrapping one single bone in the most fanciful conclusions. If they found a kernel, they invented a whole ear of corn to go with it. Krijn, Lucy, Flo: in the museums their remains assumed the form of a wild-haired child squatting beside a creek, or a timid female primate in a rocky desert.

'We dig, we find and from too little material we draw too many conclusions' – that's how John de Vos characterised his field of study. The remains of extinct hominids were so rare and lay scattered so far and wide across the globe that their finders had all the room they needed for speculation. Between one skull and the next stretched a chasm in time (sometimes hundreds of thousands of years) and space (at times an ocean lay between them). What was missing was a numerical basis, a statistical foundation.

The career of many a paleoanthropologist was based on a single find. Raymond Dart made a living from it: he received royalties from the firm of Damon & Co. in London, who made and sold casts of the Taung skull for customers as far away as Moscow. Cheerless as Dart may have seemed, he spoke of his fossil with tenderness. 'I doubt whether a parent has ever been prouder of its offspring than I was of my Taung baby that Christmas in 1924.'

During the Vietnam War and the protests against it, he had come to emphasise Taung's cruelty more and more. His child had belonged to a tribe of primates that 'seized living quarries by violence, battered them to death, tore them limb from limb, slaking their ravenous thirst with the hot blood of victims'. But could one really tell all that from a two-million-year-old fossil?

Ralph von Koenigswald, who had cherished (and rocked) his favourite Sangiran skull with equal ardour, arrived during those same years at a contrary view of humankind. 'People are brain animals,' he concluded in 1962. 'The triumph of man is the triumph of his brains.' In addition to mathematicians, philosophers and space travellers, our species had also brought forth violinists and architects. All of this thanks to the brain as the 'superior organ', which furthermore, 'in the final, decisive chapter of our development', had led to the invention of writing.

The difference between the views held by Von Koenigswald and Dart no longer had anything to do with the anatomical distinctions between the Sangiran skulls and that of the Taung Child. Skull-reading was human handiwork. The fossils of primitive humans were polished for as long as it took for them to reflect the worldview of their interpreters.

It was the American Misia Landau who convincingly demasked this tendency. During her study of paleoanthropology, she had noticed that the scenarios going around about the dawn of humankind were characterised by a striking allusiveness to the universal hero tale. In 1982, in her thesis titled *Narratives of Human Evolution*, she applied literary analysis to paleoanthropology. She dissected the most current theories about primal humans on the basis of *The Morphology of the Folktale*, the classic 'syntagmatic' approach by the Russian academic Vladimir Propp, who had laid bare the underlying pattern of epic tales. Her conclusion: the most common scenarios concerning human origins strikingly resemble, in their structure, the Russian and Scandinavian fairytales.

Pause and consider. Every reconstruction of the evolution of humans starts with the 'once upon a time' motif. The hero of the story is always a prehistoric being, a mysterious figure who lived long ago in a far, distant land. Usually a primate, shaggy-haired and primitive. One day he or she leaves their familiar surroundings (the forest) and ventures out onto the unsheltered plains (the savanna), where he or she is beset by wild beasts (lions, crocodiles). Our predecessor lacks the speed and muscle power to survive in these hostile surroundings. As in the fairytale, he or she is faced with a seemingly impossible task. The only option is to come up with a stratagem. They do this by rising up on their hind legs – in anticipation of danger. This has a fortuitous side effect: the semi-human now has their hands free. In the very next phase, they are provided with resources. Within the study of primitive hominids, this is not a magic ring or a talisman, but a stone or a stick. Our ancestors throw these at the wild animals to chase them away, until they realise that you can use these same tools to hunt those animals. Which works even better when you hack and sharpen them into spears and arrowheads.

Do paleoanthropologists actually realise that they are storytellers around a campfire, Landau wondered. Unconsciously, they fill in the gaps in their factual evidence with fictions from the tales with which they grew up.

Ever since the Greek tragedies (or otherwise, in any case, since *Don Quixote*), little has changed in the blueprint of the suspense story. But why should this be a valid template for describing how a group of apes climbs down from the trees and spreads out, walking more erectly as they go, across the grasslands?

Even an alarming conclusion like Raymond Dart's, warning us that our aggression will ultimately be our downfall, can be traced back to a literary tradition: it resembles nothing more or less than the edifying closing rhyme of the medieval farce. With her study, Misia Landau placed warning signs around the borders of every archaeological dig: 'Watch Out! Potential Pitfall!'

She helped me see things differently. In the newspaper, I read: 'Feared Viking Warrior Was a Woman'. A thousand-year-old Viking skeleton, found in a grave full of swords and daggers close to Stockholm in 1880, had recently been re-examined. 'DNA from a left incisor contained no Y-chromosome.'

A warrior buried with weapons was a man; apparently it was beyond our collective imagination to think it might be a woman. Fortunately, the DNA test had come along, like a magic charm, to expose even the most stubborn of fallacies.

NINE WEEKS AFTER sending in my cheek swab, I receive an email from *National Geographic*. The results of my DNA test are available online.

I wait until no one else is home. Then I type in my personal numeric code. The words 'Your Results' appear on the screen.

'All of us are more than the sum of our individual parts, but the results below offer one of the most fascinating . . .' Without reading any further, I scroll on to 'Your Hominin Ancestry'.

And there it is: I am 1 per cent Neanderthal. That's not much – 1.1 per cent, to be exact. The average is 1.3 per cent – but you also have highs of 5 per cent on occasion. Had I been hoping for a higher percentage? If I had scored above the average, my students would have made fun of me; the Neanderthal's image is still just plain hopeless.

Without anyone at *National Geographic* ever having seen me, they now know that I am a white male – and blond. I feel as transparent as tracing paper, even though I signed up for this myself.

Then I spot another set of results: 'Your Regional Ancestry'. My lineage during the Holocene, the last 10,000 years, is split up here into regions of origin. Seventeen per cent of my genes can be traced back to ancestors on the British Isles and Ireland, 37 per cent from Western and Central Europe, but the largest share, 46 per cent, comes from Scandinavia. I am half-Viking.

According to the explanation, I am the descendant of a group of hunters and gatherers who migrated towards the Arctic Circle after the last ice age, as the glaciers melted. My Scandinavian forefathers were from the 'historical homeland of the Vikings'. Then comes the euphemism that they, 'as a seafaring people, interacted with their neighbours in Great Britain and Central Europe'. Interaction. I'm reminded of the old history poster that hung in our classroom: 'The Norsemen anchored off Dorestad!' They had come that far, burning and pillaging, sailing up the delta tributaries all the way to modern-day Wijk bij Duurstede. It's not inconceivable that I owe my existence to a band of robbers who assaulted the local womenfolk.

At the bottom of the page, where you would expect to see 'The End', it says: 'The Story Continues'.

AND THAT HAPPENS sooner than I expect. That evening I meet up with my nephew Tom, my sister's eldest son. With a certain amount of swagger, I tell him that I am half-Viking – which makes him half-Viking too.

'But that's old news, isn't it?' he says. 'Through Grandpa, your father, we're descended from the Vikings. That's what I've been hearing my whole life.' This means nothing to me. But Tom insists. 'You've got that hand condition too, right?'

I show him my palms. Yes, Dupuytren's disease – my father and I both have it. It's an affliction, a congenital one, in which the tendons of your fingers fuse with the fascia. It makes your hand curl up like the claw of an arboreal animal. I myself have knots of tissue in the palm of my hands these days, right under my little fingers and ring fingers. My father has it worse: he's already been operated on.

'The Viking disease,' Tom says.

We look it up right then and there. And I'll be. On the site of the foundation for Dupuytren patients – I'd never heard of this foundation before – we read that, of the folk names given to Dupuytren, including 'Celtic claw', only 'Viking disease' has any relevance. The syndrome can be traced back to a mutation, a blunder in a string of As, Cs, Gs and Ts. It is particularly prevalent along the coasts (North Sea, Atlantic Ocean) and rivers (Volga, Dnieper) where the Vikings (men and women) passed by a thousand years ago during their trading and raiding forays.

I look at my cupped palms again. These scars are nothing like stigmata: the congenital defect I carry and pass along holds traces of a brutal, merciless fellow human, the signs of what one could call 'paleo-meToo'. Or am I letting myself be drawn into telling a scientific fairytale?

12

ON THE COAT RACK, bands, gown and lab coat hang beside each other in a black-and-white tableau. Once you're past the office of the prof who supervises the data laboratory, you have to take your shoes off. The entrance ritual dictates the wearing of plastic slippers – like Crocs, but without the holes.

Around the corner is the start of a salmon-pink corridor, with sample chambers on either side, reachable only through a lock. The door to the first of these is hung with radioactive warning signs: this is where the argon–argon dating of volcanic rock takes place. On the floor is a lead vat containing radioactive material; on the work surface is a Geiger counter.

José Joordens leads the way. When doing shell research, she regularly works in the radioactive zone, but what we're going to do now – the dating of shark-tooth enamel with mass spectrometry – is dangerous only for those with a pacemaker. I won't be needing latex gloves or plexiglass goggles, but a cotton lab coat is mandatory.

'Keep this buttoned up,' José says, closing the Velcro strips on my flapping lab coat.

Beyond the double doors, I make the acquaintance of a man-sized machine, consisting of three units connected by glistening tubes: the TRITON Plus Thermal Ionization Mass Spectrometer. Everything cold and clinical about this computer-operated machine is compensated for in some magical fashion by a pair of stainless-steel globes the size of soccer balls. These, apparently, are reservoirs for liquid nitrogen kept at minus 196 degrees Celsius.

Wearing surgical gloves, José empties a thermos bottle full of nitrogen into the funnel atop one of the globes; the vapours roil down the sides.

The specimens to be dated are lying in the fume cupboard along the wall: tiny pieces of ground enamel from seven fossilised shark's teeth taken from the Dubois Collection, dissolved in nitric acid. The teeth once belonged to the now-extinct giant shark – the largest ever, at eighteen metres long – known as the megalodon ('big tooth'). The enamel contains minuscule quantities of strontium phosphate. In nature, the element strontium is found partly in the form of strontium-87 (the variation with an atomic weight of 87), and the lighter strontium-86. The ratio between the two is a hidden clock that indicates how many millions of years ago the giant shark swam the oceans. The mass spectrometer can separate the strontium-87 from the strontium-86 and weigh it. In other words, it can read the clock.

José Joordens is hoping for a reading of thirty million years, but she's reckoning with the possibility of twenty million, or ten million. Or even five million.

'If it spits out thirty, then you buy,' I say.

'I'll treat the whole faculty to cake.'

The lab assistant who watches over this machine doesn't know what we're looking for. At the start of every measurement session, he places about twenty samples on a gear wheel, then inserts that through a hatch into the heart of the spectrometer. That seven of the positions are occupied by shark's teeth from Java doesn't surprise him in the least. One of the other samples in the same run turns out to have come from a *Tyrannosaurus rex* in Montana.

José prefers marine animals to land animals. 'The megalodon is the *Tyrannosaurus rex* of the seas. The biggest shark ever, with the biggest teeth. Very sexy.'

Teeth are the only thing that's ever been found of this fish, which was four times the size of the contemporary great white from *Jaws*. José got her first specimen from an Australian at a dig on Sulawesi. He himself was looking for hominids. More specifically, for the ancestor of Flores Man. For the skull-hunter, the unearthing of a megalodon tooth was nothing more than a curiosity, but José was delighted with it. Even more so when the mass spectrometer read its enamel at an age of 30.4 million years. If that Sulawesi tooth really is so old, it knocks over the existing ideas about the megalodon – including the supposition that it is no more than 23 million years old – like a set of ninepins.

In the meantime, José has sent her 'meg tooth' off by registered mail to the world's leading expert in Alabama, asking him to scan the thing and describe it for the purposes of a proposed article in *Nature*. Her (re)admission to that scientific victory platform, though, depends entirely on a solid foundation of research, which should be based on more than a single tooth. The Dubois Collection held the key: according to the list of contents, Eugène Dubois had also found megalodon teeth along the Solo River in Central Java. Seven of them. Just before the Naturalis museum was closed for renovation, José was able to fetch them from the collection tower. The question is now: is one of them also thirty million years old?

'Shark researchers are nearly as divided among themselves as paleoanthropologists,' José says, as the lab begins closing down. 'When they get together, the fur really flies.' After hours of calibration, the mass spectrometer can start in on a nocturnal measurement session all by itself. Results in forty-eight hours.

JUST AS JOSÉ is given all the room she needs to pore over shark teeth and mussel shells, her younger colleague Hanneke Meijer has all the bird bones tossed in her direction. It's like some boon from the men who are looking for hominid fossils – they leave all the fish and the birds to them.

Dr H.J.M. (Hanneke) Meijer is involved in the current excavations on Flores. Like José, she is a product of the John de Vos school of thought, and she too sees *Homo floresiensis* as a dwarfed *Homo erectus*. With her knowledge of the world of prehistoric birds, she was able in 2009 to identify four bones from Liang Bua. They turned out to belong to an extinct species of stork, a giant one, a bird whose stilt-like legs would have reached Flores Man's shoulders. Wingspan: three metres. Body weight: sixteen kilograms.

Hanneke's publication supplied fresh ammunition for the 'island rule', which serves as an explanation for both gigantism and dwarfism. The foundation for this rule is that evolution on islands is characterised by peculiar turns: some animals end up in a downward spiral, others in an upward one. On islands, deer evolve into miniatures of themselves; hedgehogs can become as large as porcupines. The possible cause behind it: suppose that the animals on a given island include elephants, but no lions or tigers. The elephant then no longer needs to be as gigantic as it is in Africa; with no natural enemies, a smaller size will suffice. Rodents like mice and rats, on the other hand, may profit from a larger body, which frees them from the need to run for cover in the face of raptors or other predators: their body temperature can then be regulated with less nutrition. They 'gigantify'.

A side effect of the island rule: birds may lose their ability to fly; they become flightless.

With her description of the now-extinct giant marabou stork of Flores, Hanneke Meijer stepped from the shadows of Liang Bua straight into the media spotlight. She related her account of the stork with so much verve that *The Guardian* promptly offered her a column of her own, under the title 'Lost Worlds Revisited'.

The idea of the fossil world as an exclusive men's club was in a state of decay. In a sort of delayed flanking offensive, female scientists were advancing into the domain of human evolution. Usually from the sidelines, or indirectly, like Hanneke and José, by way of the fish and birds. Until now, the history of paleoanthropology had been permeated with male interference. The epoch of Dubois and Dart, when wives were there to type their manuscripts, may have made way after the war for a period in which Mary and Meave Leakey carried out pioneering work, yet they remained the wives of Louis and Richard. In the debates concerning 'man's place in nature', it was the husbands who hogged the conversation. The primal human, apparently, was a 'he', in the same way God had a beard. As late as 2004, when Flo's discovery was announced, her official portraitist depicted her as a short-legged creature with male sex organs.

The striking thing was that whenever women crept out from under this male domination and gained access to the skulls of hominids, promptly there were new insights. In 2009, Dean Falk of Florida State University, who had studied the brain morphology of LB1, published her own views on what made humans human. Her emphasis on the unique human trait of language use was nothing new. But how had humans acquired this exceptional gift of speech? In her book *Finding Our Tongues*, Falk refuted the classic theory about the origins of human language (the need to communicate during hunting and warfare). That idea was the product of male brains. *Homo sapiens*, Falk said, instead owes her exceptional linguistic powers to the crying of babies. As soon as the first primates straightened up and began to walk erect, Falk's story goes, infants could no longer cling to their mothers. This meant that babies had to cry for milk. This bawling elicited from the mother a flow of comforting noises, ranging all the way from humming to speech and song (rhymes, fairytales, lullabies). This was the way the (maternal) human had distinguished herself from the animals: as mistress of linguistic ability and communication.

'The female discourse' – that was the jeering reaction, from men of course.

HANNEKE MEIJER SEEMS the most likely liaison to get me into the excavations at Liang Bua. On behalf of the University of Bergen in Norway, she is a member of the Javanese-led team that will resume digging in the spring of 2017. Six weeks of fieldwork are coming up, with excavations going deeper than ever before. Because I want to be on the spot to experience a few days of that campaign, along with my daughter, José arranges a meeting. As luck has it, we don't have to go all the way to Norway: Hanneke has spent the holiday season with her family in her native village in Brabant province, so Utrecht is convenient enough.

The moment we meet, Vera and I are struck by the fact that Hanneke wears a sweater covered in flamingos. They are all standing in somewhat different poses, balancing on one leg, neck tucked beneath a collar of down, or otherwise fully extended. We laugh.

In what is perhaps a clumsily direct fashion, I ask Hanneke how she has come to study extinct birds.

There is a long answer and a short answer, she says. We get the short one: '*Jurassic Park*.'

José comes bustling in, wrapped in scarves. We pull up our stools around a high table in Winkel van Sinkel, once Holland's very first department store, now a busy cafe. It is a blustery Saturday afternoon in January. We order mint tea with honey and hot chocolate. José has bad news about the dating of the shark's teeth: something has gone wrong with the spectrometer, so now they will have to do it all over again in Germany, at the University of Mainz. Before things get to that point, though, José hopes to get hold of more megalodon teeth, to improve the chances of finding one of the same antiquity as her Sulawesi sample.

The very next moment we move to Flores. Scooters and rice paddies pass in review, minivans with passengers on the roof, children waving along the roadside, the silhouette of a smoking volcano – as Hanneke swipes her finger across her iPad.

'How hot is it there?' Vera asks.

'Not incredibly hot,' José says. 'And on Flores, as a woman, you don't have to cover yourself from head to toe.'

'The cave is cool inside,' Hanneke says, swiping on to some photos of Liang Bua itself. We see a half-open space hooded with stalactites. The floor of the cave is clay, and the size of a theatre stage. In it are rectangular holes with bamboo ladders sticking out of them.

Men in short trousers are shovelling earth onto torn-open rice sacks. The sacks are then rigged between two bamboo poles, like a stretcher, to be carried off when full by two bearers at a time. It makes them look like paramedics lugging away the remains of Flores Man.

Hanneke shows us close-ups of the pits themselves. Their walls are of clay, sand and volcanic ash, recognisable by the colours (red, yellowish, black and grey). Each layer is marked with a thin skewer bearing a plastic card showing the geological characteristics. José calls them 'horizons', which sounds more charming than 'layers'. All the earthen material that has blown and washed into the cave over the course of time has piled up ten metres deep; its 'time depth' is determined at 95,000 years. Later, standing at the edge of one of those pits, we will stare into prehistory. At Liang Bua, the rule of thumb is: every ten centimetres you scrape off the floor takes you back in time a thousand years.

Morwood and company went wrong when they determined Flo's antiquity back in 2004. Rather than the sensationally young age of 18,000 years, including speculations that *Homo floresiensis* may have lived until the volcanic ash rains of 12,000 years ago, LB1's age has recently been adjusted upwards on the basis of the layer in which she was found.

'So how old is she, then?' Vera asks.

'Between 50,000 and 70,000 years,' Hanneke says. 'That's still awfully young, you know.'

Remarkable, though, I think: this abrupt 'ageing' by a factor of three.

'And this is the flotation sieve,' says Hanneke.

We are looking at a section of blue canvas stretched between poles, outside the cave but out of the sun; at a shady spot amid the rice paddies where the water trickling steadily from an irrigation ditch is used to wash even the smallest bones and teeth from the soil.

We see local men working with buckets, spades and sieves. They are squatting or bent over. We ask Hanneke whether it isn't awfully arduous labour.

'Oh, but I don't do that myself.' The locals see to the earthmoving, she explains, in the same way they do at excavations all over the world.

As a Dutch woman, Hanneke Meijer has never felt unwelcome in Indonesia. Proof that no old colonial grievances are being settled anymore is right there on her iPad. She shows us photos of a dark-green strongbox in Jakarta, its metal doors complete with a chromed, wheel-shaped bolt. Both doors are wide open. On a table in front of it lie plastic trays of various sizes. We see calm-looking Indonesians in shirtsleeves and Hanneke too, her mid-length hair brushed behind her ears. She is examining the contents of the trays. In one of them, in close-up and clearly visible: the original head of LB1, the only skull of the extinct *Homo floresiensis*.

WITH THE PASSING of Teuku Jacob and Mike Morwood, the controversy surrounding Flo did not simply go away. In resuming the excavations at Liang Bua, Hanneke Meijer was saddled with the legacy of their feud. In one of her columns, she provided a laconic summary of the ongoing struggle for recognition: 'Grafting new branches onto the human family tree is no bed of roses.'

José asserts that this, indeed, is not *supposed* to be easy. To keep each new skull from being declared the holotype, a constant duel took place between 'splitters' and 'lumpers'. The former wanted to give their find a separate generic name. There was glory in that.

Mike Morwood was a splitter: he succeeded in associating his name with *Homo floresiensis*. As a result of the splitters' work, professional publications contain such a plethora of branches that no one can figure it out anymore. But then along comes a lumper, who prunes away everything of shaky status, root and branch. *Homo rudolfensis. Homo pekinensis. Homo ergaster*. Once presented with great kerfuffle, then noiselessly raked together onto the heap of *Homo erectus*.

It is a pity, I think, that Hanneke and José are not more critical of Mike Morwood's opportunism.

I am reminded of Freek and Roger, who had set themselves up as the computing centre of our classroom. Whenever ciphers and graphs appeared, those two would submit them to analytical scrutiny. By running through Mike Morwood's calculations, Freek and Roger discovered that he had doctored the books with impunity (and, admittedly, openly) in order to make LB1 fit within the genus of *Homo*. Flo possessed a cranial capacity of barely 400 cubic centimetres. Her brain volume therefore lay well beneath the accepted lower limit (600 to 800 cubic centimetres) to be counted among the *Hominidae*.

'So how did Morwood solve his dilemma? By lowering the lower limit even further!'

José listens patiently to my outcry, but shows no sympathy for it. On the contrary, she stands up for Morwood. 'In the old days, a paleo-anthropologist was supposed to check off a standard list of traits,' she says. 'Skull shape, cranial size, bipedality. That's how rigid it was.' Checking off a list like that, José feels, is more the work of a customs official or tax inspector.

'Why's that?'

'Because the concept of "species" itself is a fabrication.' The outline of what is and is not a species is not a predetermined thing. In José's view, it is nothing but a concept, and a catch-all of a concept at that.

From biology class, I recall the criterion that horses and donkeys are two separate species: even though they can mate, their progeny, the mule, is sterile. Clear enough, it seems to me.

But José says that definition is obsolete. If you look at it like that, you're in danger of losing sight of the big picture, of evolution as an ongoing process. No one, after all, can ever pinpoint a fixed date at which the first human was born of an animal, or the day when *Homo erectus* crossed over into *Homo sapiens*. Before long, José predicts, the Neanderthal will be promoted to *Homo sapiens*. The differences between them, after all, are hardly more pronounced than those between, say, Maoris and Scandinavians. 'There are even those who are in favour of categorising the chimpanzee and orangutan under the genus of *Homo*.'

That, instinctively, is where I draw the line. José has me right where she wants me. Soon, she says, she will be holding a reading entitled *From Apes to Humans: A Small or a Big Step?* There is no need for me to guess what the answer – or, at least, *her* answer – will be.

HANNEKE MEIJER WAS very fond of Mike Morwood. She'd admired his recalcitrance.

'He was always a few steps ahead of us,' she tells me. 'We would be sitting around a hotel room talking about how the day had gone, and he would already be talking about going to Sulawesi.'

Even after the excavations at Liang Bua were resumed in late 2007, Morwood went on digging at the hillside with elephant remains in Central Flores. More and more flint tools were emerging from the pinkish, clotted grit, but without a single human bone among them.

'You were all expecting to find a *Homo erectus*,' José says, spurring on the conversation.

'Yes, absolutely, a big one. That's what I had my bets on.' Something like Von Koenigswald's Sangiran skulls, with a cranial capacity of around 1000 cubic centimetres – that was what they'd expected.

'But what you found was more *Homo floresiensis*.'

'That was so bizarre.' Hanneke tells us about their major victory, two years after Morwood's death. In June 2016 she wrote about their find in *The Guardian*, as one of the first. The headline above the article was 'Bones at Last', and in it she announced to the world the discovery of a piece of jawbone and a few teeth and molars so small they could only have come from Flores Man.

'A fantastic confirmation that *Homo floresiensis* was structurally small,' José says.

'Structurally weird,' is Hanneke's take on it.

The world's smallest hominid was once the norm on Flores, and for a long time too: the layer in which the jawbone was found was dated at 700,000 years. When you extend the line to LB1 (who was 50,000 to 70,000 years old), this meant that Flores Man had survived for at least 25,000 generations.

The discovery of the little, six-centimetre-long jawbone dragged Flores Man out of the twilight zone. Whether the species was now placed out on a limb or in the very crown, *Homo floresiensis* had won itself a position in the family bush of humankind.

'Apparently, though, it never dwarfed during those last 700,000 years,' I say.

But José's assumption is that *Homo erectus* had shrunk relatively soon after reaching Flores. 'Dwarfing can happen very quickly.'

That seems improbable to me. Freek and Roger had discovered

that the jawbone of the much older Flores Man found in 2016 was even smaller than the corresponding piece of jaw from the more recent LBI. 'Twenty per cent smaller, according to the scientists' own records,' they had commented. 'So what's all this about dwarfing?'

'Are you sure those figures are right?' José queries.

In a separate folder in our Dropbox, Freek and Roger had broken down their calculation to the smallest detail. I was extremely proud of them, but that isn't the point right now.

'Hmm, so the puzzle expands and keeps shifting shape,' José says, summarising the state of affairs now. Vera, too, notices that this seems to please her rather than discourage her.

I FLAG DOWN the barman, who is rinsing beer glasses, a dishtowel over one shoulder and his sleeves rolled up. Looking around the cafe, I suddenly realise how far removed our conversation is from the usual table talk here. Sitting all around us are young people who, to judge from their branded bags, have just come in from shopping. When I comment on that, Hanneke says her own hand luggage contains twenty-five fossilised duck bones from Pakistan. We look at the bag at her feet. But no – her collection is at her parents' house now.

Over cola and wine, we talk about transporting bird bones and hand axes – how customs regulations can clash with international research. José tells us more about her shark's tooth from Sulawesi, which is still in Alabama. The scan still hasn't happened, and the zoologist holding the thing has stopped replying to her emails.

José wants her megalodon tooth back. 'He's sitting on it and not letting it go.' Better no description from the zoologist, then. 'I should never have told him that I'd dated the thing at 30.4 million years.'

By now, she's assuming he is obstructing her research on purpose. But she hasn't dared to ask him to send it back to her by courier, fearing that it might then disappear without a trace.

'Are you saying that he might lose the thing on purpose, to sabotage your work?' I ask.

Before José can reply, Hanneke says: 'Some of our colleagues are capable of that. It's all about egos, right. Especially the men.'

A FEW DAYS LATER, I receive the following message:

Hi Frank,

As you know, I'm crazy about old marine stuff, I love things that come from the sea. Speaking of which, I have a strange request for you!

Could you perhaps, while the two of you are in Flores, look around to see if you can locate or photograph a fossilised megalodon tooth and, if possible, acquire a bit of tooth enamel? . . .

My Australian colleague, the one who gave me the tooth from Sulawesi, says he recognised another meg tooth in an old picture from the museum of the Mataloko seminary on Flores (Father Verhoeven's seminary!). It's quite possible that Verhoeven himself found that tooth in one of the many caves on Flores.

Could you two ask whether that tooth is still around? It would be a huge help to me!

All best!

José

13

THE MORNING AFTER the first storm of February, I drive out to FutureLand. The wind has barely died down at all, but that doesn't bother me. The tide is the most important thing; it's low water, I checked.

The navigation in the car shows the Old Maas (as the River Meuse is called here), with the villages of Maasdam and Maasdijk to my left, and to the right the Nieuwe Maas with Maassluis and Maasland. My parents, grandparents and great-grandparents all came from this delta. I have a photo of my mother as a high-school student in a striped bathing suit with leggings. She is sitting with her classmates on a wooden dock, her feet dangling above the water. 'Mr Vrieling's swimming class', says the writing on the back. During the 1948–49 school year, she learned to swim in the brackish water of the Meuse.

Since then, the wickerwork of islands and side-branches has become overgrown with industry, just like the river valley at Engis. The Shell and BP refineries are impenetrable thickets of tubing. Closer to the North Sea, the factory forest thins out, and the A15 highway narrows into the N15. Only the smokestack of the coal-fired power plant and the 65-metre-high lighthouse stick out, atop their bare poles.

On the construction sand beside FutureLand lie the now-idle carriages of the *FutureLand Express*, the tractor-driven train that from Whit weekend of 2012 was pulled across what was then the empty flats of Maasvlakte 2. Opening the door of my car, I'm welcomed by the clatter of guy lines against flagless masts. These are the windchimes of the yacht harbours. Without a whirligig or fish stand in sight.

Before the sliding doors of the visitor centre, which comprises four cylindrical huts welded together side by side, is a touring car with 'FUTURELAND EXPRESS' on the sides in light-blue lettering. After I climb out, the bus closes its doors with a loud hiss and drives off. I am a couple of seconds too late; the next tour won't begin till this afternoon – at high tide. Without a moment's hesitation I decide to follow the bus, so that later, at the beach, I can blend into the group like a stowaway.

The bus route skirts the loading platforms and the switching yards that form the outer terminus of the Betuwelijn cargo railway. We take the coastal road that runs parallel to the row of dunes and comes to a dead end in the distance. The layout of this young terrain, not yet five years old, is fixed in my mind. I look out on the quay between the Princess Amalia and Princess Alexia harbours, where the Cessna Skyhawk once dashed itself against the sand. Since the accident, humankind's technical failure has become overshadowed here by a display of technical wherewithal: fixed cranes have been built above the impact crater. They hoist containers from ships' holds and float them through the air in a choreography of stop and start.

The touring car leaves the road at beach exit number 6 and stops at a parking area atop the dune. No one gets out. Looking for fossils is part of the excursion, but the passengers seem to think it's too cold for that.

I put on my woollen cap and, while the *FutureLand Express* moves on, walk head-on into the wind to the waterline. Where the soft sand merges into hard-pack lies a ribbon of newly washed-up shells, wet and glistening. Clumps of foam break from the surf, roll onto the beach, chasing away the gulls. As I rummage about amid razor clams, cockles and plastic strips, my eye is caught by a brownish object about five centimetres long. It is not a shell, not a glob of tar, nor a cork from a fishing net.

After taking a picture of it from above, I pick it up, then rinse it off in the surf. It feels bony, softer than stone, harder than wood. One end of it is ribbed and flat, the other forked, pointy. A molar with roots and an occlusal surface?

PLEASED AS I AM with my find on this morning, I still miss my students. Or, more precisely, I miss having someone to tell about this. I am in need, in theatre jargon, of a *Sprechhund* – a dog to talk to. Even a real dog that chased the seagulls would have been better than nothing; I bet I would have called it over, proudly showed it my fossil and said something about it out loud.

But the final reportages were turned in and graded six weeks ago. We ran through them over beer, ginger tea and croquettes in a Leiden cafe. Of those who wrote a prologue, not one let reality get in their way; in one of the variations, our own skulls were pawed and classified by a team of twenty-second-century anthropologists.

The proposals for a synopsis, though, did fall within the genre of reportage. Someone envisaged a triptych consisting of the following parts:

- Out of Africa I (a description of the first probable hominid trek from Africa, some two million years ago);
- Out of Africa II (the spread of *Homo sapiens* from Africa
  across Eurasia, between 100,000 and 50,000 years ago);
- Out of Africa III (the great crossing of *Homo sapiens* in
  rubber boats across the Mediterranean Sea, from Libya
  to Italy).

'I think it would be interesting to structure the book in the form
of an excavation,' Freek wrote. 'The story would move back in time,
layer by layer.' Another student suggested following in the footsteps
of the 2.5-million-year-long journey 'from Taung Child to LB1' – from
South Africa overland to Malaysia, and island-hopping from there to
Java and Flores. That wouldn't necessarily have to take place on foot
or aboard bamboo rafts.

More feasible was Lian's 'antipode' idea: the Dutch and the
Indonesians, after all, walk the earth on opposite sides of the globe. You
only had to drill at an angle of less than one degree from a certain cafe in
Leiden to arrive at Liang Bua. Perhaps that fact might be of use to me?

OUT ON THE windy Maasvlakte, I'm reminded of my high-school
biology classes. Even though I attended a Protestant prep school, our
biology lessons depicted humankind as a creator. A cultivator of the
heath, a drainer of swamps, a digger of canals. The beaver may fell trees
to build its dams, but our biology teacher said the main difference
between animals and humans was this: animals live with nature, they
adapt to it; humans make nature adapt. We make things that weren't
there before. Including new land.

There, where Creation stopped ('Let the waters under the heaven be gathered together unto one place, and let the dry land appear'), that is where the land reclaimers and dredgers continue. The high-water line on Maasvlakte 2 is an artificial one, a man-made shore; without this intervention, the gulls would now be flying around a few kilometres inland.

I know a dredger. Ship's engineer Wim, Rotterdam accent, bald, big and strong. I run into him about once a year; he has usually just come back from places like Bilbao or Macau or Singapore. Dredgers like Wim draw the map of the world all over again – in real life. In the bay at Dubai they raised 'Palm Island': a two-kilometre-long trunk perpendicular to the coastline, with seventeen 'fronds' – a child's drawing at sea for the ultra-rich.

'We Create' is the slogan used by Van Oord, the company Wim works for. As though to underscore the creative nature of their work, the fleet of trailing suction hopper dredgers had also spat out 'The World' off the coast of Dubai: an archipelago of hundreds of islands that together form the outlines of the continents. Van Oord completed The World in January 2008; in September that same year, Lehman Brothers collapsed – and Dubai almost collapsed along with it.

'On New Year's Eve they always hold a gigantic fireworks show,' Wim told me. 'They set them off on the islands out in The World. Otherwise, they're deserted.'

Van Oord also built the beach where I am now. Over the course of four years, working seven days a week, twenty-four hours a day, the dredgers slurped up a pit in the sea floor, ten metres deep and with the surface area of one thousand football fields. With that material, they raised an embankment of sand in the midst of 17 metres of water, here at the confluent mouths of the Rhine and the Meuse, skirted by 7.5 kilometres of new beach, including 500 metres for nudists.

Amid the shells, jellyfish and jetsam lay the molars of water voles, the vertebrae of steppe bison, the anklebone of a cave lion, an ulna from a woolly mammoth and two serrated harpoon tips from the Stone Age. 'When the future is being built, the past rises to the surface,' Rotterdam's harbourmaster says.

In addition to shovelling, dredging seems another crude method for bringing archaic fauna to the surface by the bucketful. Wim has seen them boarding his dredger, the archaeologists and fossil experts. They bring strict protocols with them. For every section of new beachfront, Van Oord has to show from which corner of the concession the sand was dredged.

It was due to one of those protocols that, in 2008, the origin was determined of Holland's first (and as yet only) Neanderthal, 'Krijn'; his ten-centimetre-long fragment of skull was sucked up from a channel called the 'Middeldiep', off the coast of Zeeland province. In addition to the breakthrough of DNA, the industrial extraction of fossils from the seabed has unleashed a second revolution among those searching for the origins of humankind. Now that 'big data' is available, their field is undergoing an overhaul. The existing body of knowledge is rocked hard – and who knows, perhaps our view of humans will be rocked hard too.

BACK IN THE car, with the heating turned all the way up, I open FutureLand's 'prehistorical find checker' app. I don't want to wait for the centre's next 'bone office hours' on Saturday afternoon.

I'm asked to upload the photo of what I found. The confirmation message comes back almost right away: 'Thank you for using the prehistorical find checker. You will receive a text message from our assessor as quickly as possible.'

The object I picked up feels warm now. It's still gleaming, even after the polished surface has dried.

'Congratulations!' The promised text message comes in before I've had time to start the engine. 'This is a significant find. Where did you locate it?'

The prehistorical find checker looks like an app. But as it turns out, I'm not communicating with an AI algorithm but with a living, if anonymous, person. This service surprises and delights me.

'Beach entrance 6,' I key in, not knowing whether this information is specific enough.

'It is the molar of a wild tarpan horse. Age between 40,000 and 100,000 years. *Equus ferus ferus*. Extinct.'

14

I T WAS A FRIEND OF Wim's who infected me with the 'fossil bug': Harold Berghuis, seabed geologist. Harold and Wim had once been shipwrecked together off the coast of Borneo, in a little sounding boat that suddenly sprang a leak and capsized; they reached the coast, but it was a tough scrape. The experience forged a bond.

In 2014 and 2015 they worked together on the construction of an artificial peninsula in the harbour basin at Surabaya, Indonesia. Wim was working as a supervisor for Van Oord, Harold as an adviser to the commissioning party, the Surabaya port authority. The Javanese megalopolis on the Solo River delta was planning to expand into the sea. As a seabed geologist, Harold had gone looking for a suitable underwater quarry.

'Sand is worth money,' he tells me back home in Amsterdam. 'Especially if it's not too far off the coast and not too far below the surface.'

Harold is about as bald as Wim, but slenderer. On the landing at the top of the stairs to his apartment is a racing bike; a pair of padded trousers and a jersey are hanging from the crossbar. Past the landing, the walls are hung with modern art. Instead of his canvases, though, Harold shows me a framed navigation chart, titled 'Approaches to Surabaya'. A poster for a war movie, I think at first, but it's a publication from the local pilot service. His finger traces the depth contours of the strait that run directly past the seafront city.

'At high and low tide, the water flies through there like a bullet,' he says.

We are looking at a broad waterway connected to the Java Sea. The channel's edges are coloured a mossy green and marked with a repeated warning: 'former mined areas'. According to the map's legend, it's better not to fish there with dragnets.

At the narrowest spot in the channel lies a special shallows. Everything around it is sludge, unusable, but the submarine ridge itself consists of firm sand. 'Exactly what we needed,' Harold says. In March 2014, Van Oord sent a fleet of dredgers out to that spot; thirteen months later, a new stretch of land had been created off the Javanese coast. On Google Maps one still sees only sea; only when you switch to satellite view does a yellowish rectangle appear in the blue harbour basin. The 'Javanese Maas Flats' is what the makers call it, although the 'Solo Flats' would have been more obvious.

After the project was completed, Harold went on working for the Surabaya port authority. 'As their oil fellow.' He has an office there that he uses for one week out of every month, and he often tags a weekend onto that in order to hunt for fossils. 'I take ten litres of water with me, and ten banana crates.'

The Solo Flats lie basking idly in the tropical heat, waiting for what the future may bring. No terminal has been built yet. It is linked to the coast by a causeway, at the top of which is a guardhouse with a sliding barrier gate that opens onto untrammelled ground. Harold has a pass, but the guard would let him in anyway: the people know him, he's one of them.

In fact, Harold is a guitarist. He plays in a jazz band, but reserves the freaky solo stuff for his fossil hunting. He combs the virgin territory of the flats all on his own. To that end, Harold has divided the peninsula into plots, marked off with poles and coloured ribbons. The only road through the sand deposit is a straight strip of asphalt. In the morning he stops at each pole and leaves a crate and a one-litre bottle of water beside the road. Then he rubs suntan lotion on the back of his neck and starts picking up fossils in rows perpendicular to the asphalt. Within an hour, the bottom of the crate is usually covered in a first layer of the bones and the teeth of stegodons, hippopotamuses, primal buffalo and tapirs. When the crate's full, he carries it to the roadside, guzzles the whole bottle of water and starts in on the next plot. The harvest goes on like that till sunset – when Harold drives back down the road to pick up all the boxes.

'I've got two Komodo dragon vertebrae. Lots of turtle. Two kinds of elephant, the normal and the dwarf variety. And the osteoderms, the scales, of crocodiles. Dubois found those too.'

Harold is an autodidact. He owns a pair of sliding calipers, a magnifying glass and a shelf full of handbooks. The lion's share of what he finds he keeps at the Museum Geologi in Bandung. In colonial times, the building was the headquarters of the Department of Mines for the Dutch East Indies. In the cellar one can still find crates that belonged to Professor Von Koenigswald, who was imprisoned by the Japanese during World War II. Although presumed dead ('during the war, there were rumours that I had drowned,' he himself wrote), Ralph von Koenigswald staggered out through the gates of the prison camp in 1945, weakened and emaciated. His wife, Luitgarde, had kept his collection of molars (from *Homo erectus* and *Gigantopithecus*) in a milk bottle, and buried the most important of the fossil bones in their garden. With the exception of one hominid skull that had been on display (and which was given to Emperor Hirohito as a trophy), he managed to hold on to all his most important specimens.

These days, in Von Koenigswald's old lab, Harold Berghuis sorts and classifies his own collection: the museum staff goes out for coffee and leaves him to it. He is a man possessed.

Harold hands me a bound copy of *Neue Pithecanthropus-Funde 1936–1938*, the most-cited publication of Von Koenigswald's career. I'm allowed to flip through it. We look at black-and-white photographs of Pithecanthropus II and Pithecanthropus III, otherwise known as the Sangiran skulls, or *Homo erectus*. This is the much-contested study that Eugène Dubois defamed and disputed to his dying breath.

Dubois may have found his holotype near the village of Trinil, but Von Koenigswald's prime discoveries came from Mojokerto, southwest of Surabaya, and from Sangiran, along the headwaters of the Solo River. Harold Berghuis has his sights set on finding a jawbone or skull fragment of *Homo erectus*, but in the Solo delta.

'There's no cell coverage out on the peninsula,' he tells me. 'When I get back to town in the evening, I always call my wife and children in Amsterdam right away. So far, the first thing I always say is: "Still no *Homo*."'

Harold's ambition goes further than finding a fossil hominid: he would love to follow in the footsteps of Dubois and Von Koenigswald. To do that, for starters, he plans to get his doctorate at Leiden. With that in mind, he had made an appointment with

José Joordens at Einsteinweg 2, and took along a gym bag full of fossils.

José (I've since heard her side of the story too) was doubly aghast. When the gym bag was unzipped, the vertebrae and molars immediately took on a clandestine sheen, as though this were a shipment of drugs. That he'd been able to get these past customs!

Once the facts were all on the table, José was shocked all over again. In those loose ends she recognised a couple of dozen large animals that had all been characteristic of Trinil's prehistoric fauna. Lying there in her office, in brief, was a mirror image of the Dubois Collection. The sabre-toothed tiger was pretty much all that was missing, along with *Homo erectus*. A few other professors came for a look. How had he got hold of this? Was he really a dredging consultant in Surabaya? Was this everything, or did he have more?

Harold explained that he had collected more than six thousand fossils from the Solo Flats, but that most of them were in a cellar in Bandung.

Despite the sensation he caused, Harold couldn't find himself a PhD supervisor. Each time he tried, he ran up against the refrain: 'No context, no story.' That he had sampled the underwater sandbank off the coast of Surabaya before the dredging began made no difference. His fossils were too clean: there was no sediment stuck to them, a condition indispensable for dating. That was right, he had to admit. They had all gone through the rinsing tank of a suction dredger, after all. Harold is now thinking about trying his luck in Amsterdam, at his alma mater, the Free University.

Beside the desk in his upstairs apartment is an antique ashcan, a sort of lucky tub made of hammered copper, filled with fossils from the Solo Flats.

'The Berghuis Collection,' I say.

'Oh, no, those are the remains of the remains. I still have to classify all of that.'

Then I ask about the accusation that he's a smuggler. Don't his fossils really belong in Indonesia?

He answers my question with a question: 'You know what I find so absurd about this discussion? The Dutch used to think that Java belonged to them. Now the Indonesians think the same thing. But Java belongs to the Javan rhino.' Harold presses his hands to his chest. He says that we, all of us, wiped out the Javan rhinoceros. Or, at least, almost wiped it out. There are still about fifty of them left in a mini game preserve at the very tip of the island. 'Those fifty rhinos have a better claim to being the "oldest inhabitants" than we do.'

Picking up his iPad, he shows me his most cherished find to date: *Rhinoceros sondaicus*, or the Javan rhinoceros. He has written down the generic name himself, on the accompanying index card at the Museum Geologi. His fossil find consists of a petrified upper jaw and two molars with a layer of enamel-like veined marble. Until recently, this rhino species was also found in Vietnam. In 2004 there were still two rhinos living there, in the Cát Tiên nature reserve, a bull and a cow. But – as if in some nightmarish fairytale – they were killed by poachers. The last rhino in Vietnam was found dead in April 2011, with bullet wounds to the leg and its horn sawn off. Ground rhino horn apparently has a street value of US$100,000 per kilogram; it is sold as an aphrodisiac.

JUST AS THE BASINS of the Meuse and the Solo exhibit a major density of hominid fossils, so are the Netherlands and Indonesia among the most densely populated countries on earth. More than one thousand people per square kilometre live along their coasts and rivers. The demographer's vocabulary includes the terms 'excess births' and 'natural population growth', but one never hears tell of a 'plague of humans'. It's peculiar, really. In the same way that we suffer under locusts, the bluefin tuna suffers under us. As does the orangutan, the African antelope, the sturgeon and the nurse shark. There are 25,062 names on the Red List of endangered species; the greatest enemy of all of them is that one, prolifically multiplying, bipedal primate.

The human is an animal that has broken loose from the animal kingdom. I myself was born close to a zoo, the Noorder Dierenpark in Emmen, from which a black panther escaped on the afternoon of 15 October 1967. My mother plucked me hurriedly from the sandbox and kept me inside for the rest of the day. From the loudspeakers of the police cruisers passing by the house, we were warned that the panther, which had mauled and killed a mouflon earlier in the day, might be hiding in the municipal park.

For the animals, we are the black panthers. They too warn each other about the human on the loose – often to no avail. No other species has ever succeeded in slipping out of nature's cage. Since then, we have spread out over the globe like football hooligans knocking down a barrier and rushing onto the field. Within a lifetime – my own – the world's population has doubled (from 3 billion to more than 7 billion). Some 200,000 new specimens arrive every day.

Darwin remarked back in the nineteenth century that this 'most dominant of animals' was manifesting its 'immense superiority' around the globe. Three-quarters of a century later, in 1961, the geneticist Theodosius Dobzhansky warned against the rampancy with which the human swarm was colonising the planet. 'Man is the most successful product of evolution, by any reasonable definition of biological success,' he wrote in his classic essay 'Man and Natural Selection'. 'Man began his career as a rare animal, living somewhere in the tropics or subtropics of the Old World, probably in Africa. From this obscure beginning, mankind multiplied to become one of the most numerous mammals . . . He is well on the way towards control or elimination of the predators and parasites which used to prey on him.' Dobzhansky stressed that we as a species have galloped beyond any point of return.

But what exactly unchained *Homo sapiens*? When did our ancestors first fall out of step?

In Classroom 0.04 we'd talked about the Great Leap Forward – the term biologists use to refer (in highly Maoist fashion) to the moment when *Homo sapiens* began pulling away quickly from all the other species. That happened 40,000 to 50,000 years ago. 'Leap' didn't sound like the right word to us, nor did 'forward'. It seemed much more like a distancing, a wandering away from the herd, followed by a tumble, a fall from nature. Within a short period, an explosion of cultural expressions – ritual, artisanal, artistic, culinary – took place. What brought this about remained unclear. Some researchers ascribed it to the development of the larynx, others to a mutation in the 'language gene' FOXP2 (also present in the Neanderthal and the zebra finch), yet others to the preparation of food using heat ('how cooking made us human'). A fourth theory accredited it to the knock-on effect of successfully passing along knowledge and skills. Whatever the cause, *Homo sapiens* took off. And was the only one to do so.

The solitary trajectory of *Homo sapiens* came at the cost of the world's other species. This had already started with our predecessors, who hunted the tarpan horse into extinction. As a silent witness of this, during a February storm in 2017, yet another molar of this *Equus ferus ferus* washed up on the beach of the Maasvlakte. Not a pretty thing.

If you ask me, humans are best typified as 'un-animals'. Beavers build dams, termites build mounds: we build hydrogen bombs. That we are just as good at making things as we are at breaking them fascinates me. Someone carves images of Buddha in the face of an Afghan mountain, 55 and 35 metres tall; someone else comes along and blows them up. Every act of construction is followed by destruction, every destruction by tireless construction. The process of creating and destroying, over and over again, keeps us going, at least, like mice on a treadmill.

IN THE SPRING of 2017, just in time for the new tourist season, the Van Oord crew is pumping twenty kilometres of new beach onto the east coast of England. This is the fourth year in a row, each time along the same stretch in the county of Lincolnshire; during the yearly winter storms, the beach is washed away.

I'm allowed to go along, on board the *HAM316*, a 10,000-tonne steel juggernaut that sucks up the sea floor nineteen kilometres offshore. The day is clear and cloudless, but the *HAM316* is nowhere on the horizon. Our shuttle, the *Offshore Phantom*, makes a beeline for it nonetheless – purely by means of GPS signals.

The radar keeps an eye on the shipping lanes, while the sonar monitors the sea floor. Directly beneath the keel, the North Sea is twelve to fourteen metres deep. There's not much swell and heave, but the shuttle leaves a swirling, foamy V in its wake. The brick lighthouse of Grimsby recedes in the distance.

'This is *Offshore Phantom* for *HAM316*.' Our captain holds the mouthpiece of the marine telephone to his stubbly chin. 'We've got three pallets of provisions on board.'

The *HAM316*, one of Van Oord's flagships, answers with a swell of static. Then we hear: 'Okay. And how many pax?'

'Three pax as well.'

The 'pax', the human cargo – that's us: dredging engineer Wim, superintendent Gerald and me. Wim is coming along to install new software. Gerald, a thirty-something from Brighton, is returning from shore leave. I'm here for the fossils.

'They're waiting for you there,' Gerald says. 'Washed and sand-blasted.' At portside he has bought tabloid newspapers for his men, all of them equally pro-Brexit. Later, on deck, they warn me that I mustn't start on about how the British Isles were once shackled to continental Europe, 7000 to 8000 years ago. Workfolk tend to associate factual knowledge with pedantry.

In the wheelhouse, I see that the *Offshore Phantom* is moving steadily away from the Dogger Bank. The *HAM316*'s collection area lies further south. That's too bad, because these shallows, where the herring spawn and fishermen have in the past pulled up evidence of human habitation, are a submarine potter's field. As the polar icecaps melted, it seems the Dogger Bank was the last place to be flooded, and so served as refuge for humans and animals.

The *Offshore Phantom* re-establishes contact with the *HAM316*. 'Twenty minutes to go.'

The captain of the *HAM316* reports that the dredging must go on without a minute's interruption. He's willing to take the pax on board right away, but the *Offshore Phantom* will have to wait to bring on the pallets until they've hooked up to the 'sink line'.

'That sink line,' Wim says, 'comes directly from Maasvlakte 2.'

I have no idea what a sink line is.

Alongside now, our boat is dwarfed by the giant ship. Seizing the rope ladder (you wait till the swells lift you as high as possible, then climb fast), we scramble like pirates up against the hull and over the railing of the mothership. There are only men working here, two dozen of them, but the ship herself is a lady. *She.*

The ranks on board run vertically: the highest position is seated in a swivel chair on a raised platform in the middle of the bridge. We are granted an audience with the captain, as long as we take our boots off first. The length and breadth of the bridge is covered in carpet: we shuffle across it in our socks. Past a black sofa, we arrive at the coffee corner.

I ask the captain how long he's been knocking about the world of dredging – 'a long time' – and what the biggest change is that he's seen during that time. Staring through the glass wall at the helmeted men in orange passing through the gangways, he says: 'That these guys with the big biceps aren't allowed to touch anything anymore.'

Behind the lounge corner with the leatherette couches is a computer with two monitors. They show valves and hatches, all interconnected. The controller on duty can open or close them by hand, but he's not supposed to. What he's supposed to do is not touch anything. So what are they watching for? 'Fossils?' I venture.

Laughter. 'We had an archaeologist in here one time,' the captain says. 'The guy looked about as stale as the story he told.'

'UXOs,' the controller says. 'That's what we watch for.'

UXO, as it turns out, stands for 'unexploded ordnance': duds, ammunition boxes, mines.

Wim tells me about an accident that occurred while they were building the Solo Flats: one of the ships slurped up a seventy-year-old mine, and the thing exploded inside the cutter head. It didn't blow a hole in the hull, but it did knock the engines out of alignment. 'Fixing that cost us millions.'

As I listen, I see in my mind's eye the navigation chart at Harold Berghuis's place, 'Approaches to Surabaya', with the text 'FORMER MINED AREAS' marking the moss-green edges of the channel. Those minefields were laid out by the Royal Netherlands East Indies Army, in a fruitless attempt to deflect the Japanese invasion of Indonesia.

STAIRCASES PAINTED IN red-lead primer lead us down through the hierarchy of the *HAM316*. Past the officers' mess we end up in the mess hall for the Filipino crew-workers, then we follow a corridor lined with single cabins and arrive at the ones with bunks. One level down is a gymnasium, with pinups on the walls by way of inspiration. The canteen is an old brown cafe with panelled walls and dredging trophies like ammo casings and the spotted shell of a giant turtle. The steel walls around us are atremble. The engine room is filled with bellows and roars; you can't go in there without earplugs.

Outside, on the rear deck, 129 metres from the prow, is a platform that serves as a resting place for the draghead when no dredging is going on. The massive thing hangs from a tubular suspension arm – think of it as a vacuum hose. We're allowed to get close enough to look into the head. It drips and splutters. Before the dark hole of the suction mouth is a steel grid. If there is one place where large fossils could become stuck, this is it. But the grille holds only a boulder. No tusk, no antlers, no pelvis of an antique hominid.

Closer to the fount of seafloor fossils I cannot come.

Later that day, I will get a second chance.

From the Lincolnshire shore, Wim and I watch Chapel Beach

broaden visibly. The waterline is steadily receding; each new wave
strikes a metre or two further away than the one before. That's because
of the river of dredged sand pouring out of the sink line. In the time it
took the *Offshore Phantom* to drop us back on dry land, the *HAM316*
has come in to sit just beyond the line of surf. Only now do I understand the function of the sink line: it is a 200-metre-long pipeline
positioned perpendicularly to the beach; when it's hooked up to the
dredger's hopper, the discharge can begin.

'We can only rainbow the sand onto the beach,' Wim says, 'when
we can get in closer.'

The sink line's nozzle is a rubber 'swing bag': a flexible section of
tubing that lies walrus-like on Chapel Beach. Bulldozers – 'fitted out
in the Van Oord club colours' – have jammed this spraying mouth in
between two mounds of sand, to make sure the watery blubber glops
out in the right direction. Within forty-five minutes, 16,000 cubic
metres of new beach arises.

Meanwhile, the bulldozers are busy spreading the newly arrived
sand. They plough back and forth, blowing clouds of steam, wading
up to their axles in the sludge. Their blades never stop moving. Raised
first when pushing a too-high wall of sand, then sinking back down
again to shovel up an even greater load of muddy mass. It's a ballet
that I'm watching. A dance concert held by two gargantuan tracked
vehicles on a North Sea beach.

I praise the drivers, and Wim shifts his weight from one leg to the
other. 'You see those white caps on both side of the blade?'

Those are antennae, it turns out, that Wim has connected to GPS
software. The shape of the new beach is programmed. The movements
of the bulldozers' blades are prompted by a computer; the drivers
themselves do nothing but go forwards or backwards, for as long as

it takes to make the surface they're moving across match the desired profile.

When the spraying of sediment stops, the bulldozers go on for another ten minutes. Then it's lunchtime.

I am the first person allowed to set foot on the new Chapel Beach. The sand is still loose and damp. Except for a few pebbles and shells, I find nothing special – unless perhaps it's the two fish I see lying in a bulldozer's track. Plaice, dab and flounder, I realise, are all flatfish that can't get away fast enough. They disappear through the draghead into the belly of the ship, only to be spat out a few hours later, on their backs, atop a new sandbank.

I have less luck on Chapel Beach than I did on the far shore, at Mammoet Beach, in Futureland. Or maybe I should have gone looking further south, in Norfolk, where 800,000-year-old footprints and hand axes belonging to *Homo antecessor* have been found, the 'pioneer human' who was the first to explore what is currently Europe.

In the workmen's hut, I talk to the drivers. They're watching news about Brexit.

'Any fossils?'

'Nah, haven't seen anything yet.'

'It's not like we're really waiting for them, you know.'

'You know how much it costs to park these machines off to one side?'

'But if a pot of gold turned up, well, then we'd stop for a bit.'

WIM TAKES ME aside. He says he didn't want to start on about it while we were on board the *HAM316*. And not with the bulldozer drivers around either. We head up towards the path that cuts across the dunes. The first tourists of the season have crowded in beside the chain-link fences around the dredging site. Men, women and children, eating from paper bags.

'There's no stopping computerisation,' Wim says. 'Lots of dredgers are worried about their jobs, but we've already started thinking about the future.'

I'm distracted by the sight of trailers, stretching out as far as the eye can see. Rectangular, leaden-coloured, in serried ranks. At their far extremity is an amusement park by the name of Animal Farm. There are stalls selling fish and chips and cotton candy, and fruit machines and whirligigs – and amid all that a caterpillar-like stream of bipeds.

Wim goes on talking. He's part of a think-tank that's designing the '*Vox Futura*', a ship of the future that should be operational by 2038. The idea is that it will be a 'self-dredging' vessel. He paints for me the picture of a floating 3D printer that can spew out sandcastles in any shape desired. Unmanned.

From atop the dune, we look out over the sea. The *HAM316* has already disappeared from sight, behind a wind farm. Rising up on the horizon is an imaginary ship that comes close to the pinnacle of human construction. The design for an elegant, tulip-shaped island off the coast of the province of South Holland is already finished.

I can't help but be amused at the thought of a species that falls from nature, sullies and distorts it, then desperately attempts to make itself redundant.

15

THE VOLCANIC CONES of Java file past us, one by one. Some are poking their heads through a ruff of foamy cloud. They look like flightless birds with unsightly, puckered necks. With the help of the flight-path map on the seatback screen, I'm able to locate Mount Merapi and Mount Merbabu. For hours we follow the Pacific Ring of Fire, but my daughter would rather watch movies: fire-breathing mountains frighten her, as does flying itself.

Ever since our transfer at Jakarta, it's felt as though we're skimming over the earth's crust like a stone skipped across the water. The longest leg, from Schiphol Airport to Sukarno-Hatta International, is behind us; now we're cutting a second arc to Bali, after which comes the final flight to Flores. A hop, a skip and a jump.

Meanwhile, the landscape below has grown greener and flatter. I pick out the Solo River, loop after loop – no serpent ever twisted so wildly. To our right and below must be Sangiran; ten minutes further lies Trinil. We are following the downstream course of 'the cloaca of Central Java' (Von Koenigswald), which fills during the rainy season with 'lukewarm and murky muck'. Half an hour later, Surabaya slides into the oval frame of my window: a clump of stone on the coast with tendrils of grime fanning out into the blue of the Java Sea. I see the yellowish rectangle of the Solo Flats, but I can hardly share my excitement. Looking straight down horrifies Vera even more.

The most sensational news of the week I also keep to myself. On Sulawesi, a missing rubber tapper has been found: as yet undigested, inside the intestines of a gigantic python.

Photos of the find have spattered all over the world. You see a dead, battered snakelike object stretched out on the ground, with a lump between head and tail. If you were to make a drawing of it, adults might think it was a hat. Standing around the dead animal is a group of excited villagers armed with clubs and machetes. In the next shot, one of them is neatly peeling open the fishy skin. The python is being unzipped like a pegged skirt, to free the body of a 25-year-old fellow who went off into the woods in the morning and never returned. He is still wearing his boots. Rubber boots.

I'm reminded of my students, who accused me of sensationalism. Whether it's the Mormons' polygamy or Miss Keers' blood tests among the Negritos of Flores, I happen to be drawn to extreme stories. That's true, I said; you're right. I tried to justify my predilection for rarities. 'If these are exceptions,' I said, 'then don't they obviously shed light on the rule as well?'

That question was met with scepticism. Objections were raised to my approach ('not very scientific'). Why linger over excesses when you're out to discover how things *normally* go?

I seized upon the criticism as an opportunity to explain why I, right after graduation, had traded in the groves of academia for the world of journalism. The scientists' disdain for the sensational annoyed me, since I myself was highly attracted to it. The more improbable

the event, the more it put me on the edge of my seat. By way of example – as an illustration of the phenomenon of the 'cliffhanger' – I told
them about a woman who, while skydiving above the island of Texel,
pulled her ripcord prematurely. A corner of the parachute caught on
the plane's tail. As a result, the skydiver was yanked horizontal and
dragged across the sky like a human ad banner. Meanwhile, the plane's
fuel gauge was nodding steadily towards empty.

*Lots of things that make the news almost never happen. That is why
they are news.* To this basic rule, borrowed from the website of the
Dutch evening news for children, I added that uncommon things nevertheless happen frequently enough to fill the news bulletins each day.

Scientists are on the lookout for averages, for patterns. Anyone
wanting to know the speed of fall of a parachute is looking for a
number between 0 and 1, the friction coefficient – and they don't
give a hoot about a freak occurrence like the one in the sky over
Texel. Science is seeking the general, not the personal; the universal, not the singular.

In my view, there were two things that justified my revelling in
the exceptional. The first was based in empiricism: it is consistently
the factor of the unexpected that alters the course of history. Human
activity is characterised by whimsy. The second: if I succeeded in accurately clipping the exception out of reality, didn't that automatically
take us beyond the contours of the rule?

WE CAN ALMOST reach out and touch the Komodo islands, with
their white sand and spoon-shaped bays. At only a little distance from
the nature preserve for the islands' great reptiles, our prop plane just
squeaks through its landing on Flores. Alfred Russel Wallace never

set foot here; he merely sailed past. 'Flores is probably one of the most difficult and unpleasant islands to travel across in the entire archipelago,' wrote his assistant in 1892 after going ashore for a look. 'There are no roads . . . and it is extremely difficult to acquire victuals and supplies.'

Extreme desolation like that on Flores at that time must have been what Eugène Dubois had in mind in 1888, when he presented his search for the missing link during a reading on Sumatra: 'This transitional form may have remained in place for a long time . . . in the Malay landscape, where all that existed was the necessary isolation.' More than a century later, in 2005, *Nature* expanded on this assumption in an editorial, saying that the discovery of Flores Man only increased the plausibility of folktales about mythical, humanlike figures being based on actual reality.

On the tropically hot tarmac at the bottom of the stairs, we are treated to two male Komodo dragons: many times life-size, slimy jaws open wide in a duel to the finish, using their scaly tails for support, they decorate the front of the arrivals hall at 'Komodo Airport'. But we are not here for existent reptiles; we are here for deceased hominids.

Waiting beside the baggage carousel, I switch off my phone's airplane mode. A whole row of text messages immediately dribbles in, all from Jeanette van Oostrum, a former classmate at the Assen Protestant School Community. For the last three years, Jeanette has been running a bed and breakfast in Central Flores, in a kampong on the south coast by the name of Leko Lembo, where we are expected in two days' time.

Leko Lembo is to be our base of operations, and Jeanette our hostess and fixer. She is the youngest daughter of Pastor Van Oostrum, who was our high-school religion teacher. In class, as he fixed his gaze on us from beneath his bushy white eyebrows, his favourite subject had been the years he'd spent as a missionary on Sumba, the island just south of Flores. Jeanette was born on Sumba in 1965 and, after wanderings that lasted half a lifetime, she'd settled down at last beside a gravel beach on Flores.

Once upon a time, in our last year at school, we had slow-danced and flirted, although after that we'd gone our separate ways: she headed back to Indonesia where, with her flair and her fluent Malay, she worked as a guide on adventure tours. In April 2014 she married an islander ('a Flores man', I told Vera). I had received a wedding announcement from 'Sius and Janet'. She had shortened her name. Together they built a house with guest quarters and a verandah from which Janet, in clear weather, could look out at the palmy coastline of her native island.

I read the text messages aloud to my daughter.

Today I heard from our household help, Efi, that there are plenty of 'little people' still out in the forest.

There is a spot 15 kilometres from here where they apparently show up often. They come there to eat rice.

Before we had even left Holland, Janet told me in an email about a lady in her village who sells palm wine and who regularly sees those creatures crossing the road at sunset.

When you two get here, I'll ask Thom, my uncle, to tell you more about them. I've invited him to have dinner with us the first night.

AT THE FAR end of the 'Nothing to Declare' lane, Philippus is waiting. Jeans, t-shirt, deep-set eyes. He comes up to my shoulder; he's sinewy and slim. From under his chin hangs a curtain of black hair. It's hard for me to tell his age – is he older than me or younger?

By way of greeting, he tosses us a handful of Dutch words. We can make out '*hoe gaat het*', '*welkom*' and '*zeker moe*'. Sure, we can do a little '*knikkebollen*' – catch forty winks – in his car later on. This is hard for me to place. Are these echoes of the colonial past?

Our driver giggles. Like Janet, he was born in 1965, as it turns out, two decades after Indonesia declared independence. He owes his Dutch vocabulary not to colonialism but to tourism.

Janet has sent us Philippus as he's the best interpreter/driver she knows on Flores, a person who can show you more than just Lonely Planet's top five things to do. Still, we deliver a blow to his professional pride by turning down the boat trip to the remote beaches where Komodo dragons abound. We want to get away from the coast, into the interior, in the hope of reaching Liang Bua by tomorrow. We're in a hurry: the excavation season is coming to a close. Hanneke left Flores yesterday from this same airport, on her way to Australia for a convention in Canberra.

Within the first kilometre we make a pitstop for a crate of bottled water at Labuan Bajo, a tropical bay in the shape of an amphitheatre, full of fishing boats and with vine-covered cones of rock protruding from its waters. Philippus parks his Toyota beside a traffic sign I can't quite figure out. The orange sign shows a little man running up a slope, pursued by a monstrously huge wave that is on the point of cresting.

'Those are new,' Philippus says when he comes back with our water supply. 'They're tsunami warnings.'

As far as Vera's concerned, it's time to head for higher ground. Past the centre of the little port town of Labuan Bajo comes a flat stretch with square and rectangular rice paddies; after that, the island's only through road climbs via a few curves into the jungle.

I ask Philippus about the 'little forest people', but that's a non-starter: he starts whistling through his teeth, as though I'm telling a joke.

'*Ebu Gogo*?' I try.

Phrases like 'never heard of them' or 'beats me' are not part of the repertoire of the world's hospitable peoples.

Maybe I'm talking about monkeys? Philippus pulls over to a roadside stand and, without climbing out, buys a bunch of bananas. 'If we see them, we feed them.'

Flores is home to both wild and tame macaques. And we should feel free to call him Philip.

I ask if that's what his friends and family call him too.

'At home they call me Lippus.'

THE TRANS-FLORES HIGHWAY is a barely two-lane blacktop that winds more wildly than the Solo River. After three hours of driving, by the time Lippus suggests we go to see the celebrated wheel-shaped *sawas* of the Manggarai, we're feeling empty and dizzy. The stop is a welcome one, even more so when it turns out there is a stand selling cold soft drinks at the foot of the lookout point.

The Manggarai plant their rice in circular parcels; each family owns an irrigated slice of the pie. The fields lie tucked in amid steep, wooded slopes. In the distance, two water buffalos, a big one and a little one, are taking a mud bath.

I have to goad Lippus into passing along my question about the little people to the couple running the drinks stand. Their daughter fills in for them as they walk with us to the shade of a thatch-roofed arbour.

Lippus is stammering. I can tell by his gestures that he's apologising beforehand.

'*Ebu Gogo*,' I interject.

'*Bibet!*' the woman says, her cheeks puffing with laughter.

'*Bibet*,' her husband says as well. The next moment they ask if I'd like to take pictures of the *bibet*.

'Yes,' I say, 'I'd love to.'

'That's not too easy, you know,' the man says.

I turn to Lippus: *Come on, help me out – what's a* bibet?

Lippus has no idea.

But he can ask, right?

'It's Manggarai, and I speak Ngadha,' he says.

Come on, Lippus.

After some fumbling around, the translation comes in. *Bibet*, it turns out, means something like 'sprout' or 'seedling'. If you're cultivating potatoes, you put aside some seed potatoes after the harvest: *bibet*. If you're planting corn, then you keep a few ears for the following season: *bibet*.

These sprouts are hairy little people. They live in the forest. 'About this big' – the man holds out his hand at the height of an upright bale of rice.

'Have you ever seen them?'

Laughter. Of course! We're treated to an imitation right away: the soft-drink salesman crouches down and holds his hands up like claws next to his ears.

'What would I have to do to get a picture of them?'

'Dad!' From the corner of my eye, I see that my daughter's trying to get my attention.

'Get up early and go into the woods with the hunters.'

'You never get a clear picture of a *bibet*, though,' the woman warns me. 'As soon as you click the shutter, they go racing off.'

Vera is sitting beside a boy of about fourteen. '*Whasyourname, whasyourname?*' he asks, shaking her hand. 'Dad, he won't let go,' she pleads. 'I've already told him my name ten times. What am I supposed to do?'

I hold out my hand to the boy and introduce myself, and grab his at the same time.

A little later, back in the car, I hear Vera sigh loudly from the back seat: 'Culture shock, here I come.'

I HAVE MANOLA'S final report in my bag. It starts with a story about the Iné Wéu of West Flores, a half-human, half-animal denizen of the jungle. She is related to the *Ebu Gogo* of Central Flores, although, Manola says in a footnote, Iné Wéu and *Ebu Gogo* may also be different names for the same, arguably mythical creature. Father Jilis Verheijen of Steyl, in a church publication in 1978, gave the following description:

> It is above all in the forest that people are afraid of Iné Wéu. When encountering this creature, who captures humans with her pendulous breasts, it is important to have a dog with you. In more or less historical accounts, the usual course of events is that a person so caught has the momentary opportunity to pinch the dog, which he has brought with him in a basket. The ensuing yelps cause the Iné Wéu to flee.

With the help of her father and three dictionaries, Manola succeeded in beating a path into the original prose that Verheijen had collected. In italics, she recorded lines taken from Part V of the *Manggarai Texts*, page 491. Beneath that came the literal translation, and beneath that, in upper-case letters, 'What I think we're able to make of it'.

> 1. *Bo Iné-Wého'o, cama ge ela kinay taran.*
> *Awhile, feral spectre/ghost was there, like unto pig his/her, mother animal, frame.*
> SOME TIME AGO THERE WAS A FERAL GHOST THAT LOOKED LIKE A SOW.

> 2. *Lémpa nggeréta békék o'os pe cucun, émé lako hia.*
> *Something over the shoulder hanging, it is so, this his/her breast, if/when he/she walks.*
> SHE CARRIES HER BREASTS OVER HER SHOULDERS WHEN SHE WALKS.

Lines three and four led to the following translation:

> THE APE-PERSON RETURNED WITH ANOTHER APE-PERSON. WHEN SHE DESIRES HUMAN FOOD, ALL SHE HAS TO DO IS BEAT WITH HER BREASTS IN ORDER FOR PEOPLE TO GIVE HER THE FOOD SHE DESIRES.

Lippus explains how divergent the languages of Flores are. The wooded mountains we're crossing now are part of the Manggarai region. Barely a kilometre past Janet's guesthouse in Leko Lembo begins the more arid Ngadha, where Lippus comes from, with its thistles, lava beds and smoking volcanoes.

As far as the similarities go between Manggarai and Ngadha: of everything that has just been said, he recognised only the word *ata*: 'person'. *Kita ata*, in Ngadha, he translates as 'human being'. *Kita* means 'we'. *Ata* is 'people'. *We, the people* is what we make of it – that sounds less rigid and formal than *Homo sapiens*.

A little later, I ask Lippus how he became a chauffeur for tourists.

'I wanted to join the army, become a professional soldier,' he says, sounding wistful, 'but I was too little and too skinny and my family didn't have enough money to bribe the right people.'

So he had gone to Timor to work as a conductor on a minibus run by a Chinese family: standing on the running board, hanging out of the open sliding door to fish passengers from the shoulder of the road, grasping a fan of folded rupia notes between his fingers by way of a wallet. He was good at it.

But then his sister got married, and Lippus came back to Flores for the wedding. 'In two weeks, I spent all my savings.' At the wedding he met a driver who took tourists around the island. All you had to be able to do was speak English – that was all. 'You just talk' – that's how you earned your keep. It was incredible, but true. In the meantime, he can also welcome people (and swear) in French, German, Russian, Italian and Dutch; he's been doing this for seventeen years already. But he got off to a difficult start. Lippus tells us that he had three accidents during the first year. His second-hand Toyota Kijang, as it turns out, is not his property. His aunts and uncles, nephews and nieces all have shares in it.

'My family took me to a shaman back then.' The shaman locked him up in the communal longhouse, shrouded with incense. The conclusion was that there was some conflict with two of his ancestors, a fat one and a skinny one. The remedy prescribed was that each year, in the company of the shaman, Lippus was to present an offering of incense to them. This was more effective than a costly insurance policy: he has been driving without an accident for the last sixteen years.

When I question him further about the shaman, Lippus leads us into an unfamiliar world. If a child is choking on a fishbone, there's no need to go to the doctor. All you have to do is call in someone who was born feet-first. Those who come into the world by a breech birth possess supernatural powers.

The story about the *bibet* has seemingly stirred up a memory of what Lippus's mother used to call the *koersjati*. Those are *orang pendek* – 'little people', in Bahasa Indonesia – and you have to watch out for them.

'You should never feed them,' he recalls.

'Why not?'

'If you see them, you either have major good fortune, or major bad luck, right away.'

'What kind of bad luck?'

'That a driver coming the other way knocks the mirror off your car in the next bend, for example.'

Now that the *koersjati* are fresh in his memory again, I ask whether he believes in them.

'Ay-ay-ay,' he cries. 'You know what it is . . .' Lippus doesn't complete his sentence, he plucks at the hair under his chin and starts anew. 'It's just like with Jesus. It's hearsay, a story that makes the rounds, without you knowing anyone who's ever really seen him.'

AFTER SUNSET THAT EVENING, at Hotel Sindha, where LBJ once lay in state in Room 109, we've got internet access again. Reception is best amid the plastic potted plants in the harshly lit dining room. Janet texts: '!!!', followed by a link to the UK daily *The Sun*:

> *Canberra, 22 April 2017 – Flores Man – 'Hobbits' – Not Direct Relatives of Modern Humans*

At the opening of a conference in Australia, Dr Debbie Argue cornered the collected media with her findings.

> *Experts initially believed the* Homo Floresiensis – *or Flores man – were just a shrunken variety of early humans.*

But that's not true at all, according to a publication in the latest issue of the *Journal of Human Evolution*. In the accompanying photo, Dr Argue – blue sweater, blue-framed spectacles – is holding a replica of the LB1 skull in the palm of her hand. They are looking at each other. 'Flo' is not a dwarfed *Homo erectus*. 'On the family tree of the genus *Homo*, there is no way they can be placed on the same branch,' she says.

Argue's analysis of 3D scans of hominid skulls ostensibly puts paid to the ideas of John de Vos, José Joordens and Hanneke Meijer – in short, 'the Dutch school'. I feel like contacting them right away: Hanneke is in Australia at this very moment, so she feels close by. Does this constitute the coup de grâce for the dwarfism theory, from which Mike Morwood too disassociated himself at the end of his life?

I have a little trouble falling asleep that night. Is this the dialectic–thesis–antithesis–synthesis needed to arrive at knowledge, or is it only a few egos (male *and* female) fighting each other over a bone? And another thing: why do newspapers adopt such pompous names, like *The Standard* or *The Truth*? Or, even more presumptuous, *The Sun*?

16

OUR APPROACH TO Liang Bua the next morning turns into an impromptu motorcade. For forty-five minutes we drive past long rows of houses stretched out along a ridge. The road is bumpy and unpaved, but what Lippus must pay special attention to is the sizeable crowd we're drawing. Whenever someone catches a glimpse of Vera's light-blonde hair, both children and adults come racing out to the roadside. Waving and grinning broadly, Lippus honks, Vera and I wave back uneasily.

Then comes the downhill stretch through a banana plantation. The Toyota slips and slides its way along the descent, lurching like a mule. It is eight o'clock; the sun comes up from behind a mountain wall and pours wagonloads of light onto the forest canopy. Here there are no hedgerows of curious Floresians, only tree ferns and other luxuriant growth. We inhale the moist air of decay.

In the valley below, two brooks converge. A few million years before we got here, one of them, Wae Racang, carved a hollow in the limestone wall and then filled it up again with layers of sludge. 'Selamat Datang / Welcome at Liang Bua'. On the far side of Wae Racang, the road is spanned by a triumphal arch of concrete and steel. We are driving between the rice paddies now, without a cave in sight. Yet here, too, boys and girls in flip-flops come running out to greet or examine us.

It is a bizarre sensation, to visit an attraction as an attraction oneself. We have to walk the final bit – under escort – uphill to the forest's edge. Halfway up this path we're joined by a line of workers – dozens of them. As though at some film director's cue, they come traipsing up in single file over a narrow dyke, each carrying a bucket on one shoulder. Between the rice paddies lies an uncultivated patch; that's where they're coming from.

The flotation sieve! Hanneke showed this to us last winter, on her iPad in Utrecht. Through a canvas tarp, down there by the brook, the smallest bits and pieces of teeth and bone are separated from material taken from the floor of Liang Bua. But now the men are actually climbing *uphill*, with full buckets.

'This is backfill,' their foreman tells me.

It takes a moment for me to understand what's going on here: his brigade is carrying the sieved mud back up to the cave, to fill the pits dug during the last campaign and make sure no one can fall into them. Then we notice that all the bearers are wearing identical T-shirts, grey ones with 'HOBBIT' written in yellow on the front.

I want to take their picture, which is easy enough, as no one here ever says no. One of the bearers swings his bucket down onto the ground and squats beside it. 'Human mysteries in Flores' is written in smaller letters beneath 'HOBBIT'. The man scowls.

Lippus comes to bring us bad news. 'Liang Bua is closed to visitors,' he says. There's work in progress ... and besides that, we should actually pay something for the pictures we just took. One cigarette would be enough.

But I don't smoke.

He asks me for three 10,000 rupiah notes, then he takes the foreman aside and makes a deal. A messenger will be sent running to the village to buy two packs of *kreteks*, the local clove cigarettes: one pack for the foreman and one for his crew. Liang Bua, its entrance hidden behind huge taro leaves, is then open to us.

We enter one by one, Vera leading the way. It all looks different from the photos. The light is more diffuse inside, less abundant. Tree ferns hang like a whale's baleens from the top edge of this gaping hole in the earth. We descend a few steps to a clay floor beneath a palate of stalactites. Down here, amid a few upright spikes of rock, men are at work with spades. Piles of earth to be shovelled back into the pits lie here and there, but I can't escape the association with birds picking clean the teeth of some carnivorous creature. We have walked into the maw of a colossal reptile.

Now the bearers come in again as well, some of them with lit clove cigarettes in their mouths. The tips glowing and fading in the dusk, they file in an ants' path to the trench and dump their buckets into it.

We've arrived a day too late. Or have we? The filling of the pits began only this morning; fortunately, the biggest one is still empty. Looking over the edge, we see diverse layers of dried sludge, each with its own colour and texture. Little cards have been stuck into them. 'Sandy silt mixed with limestone.' One card marks the layer of the volcanic ash that incinerated the treetops of Flores 12,000 years ago, along with all the creatures, great and small, living beneath them.

The groundwork crew takes a break. From knapsacks hanging on sticks in the cave wall, the men produce their spring rolls. The foreman, 'John', also wearing a grey-and-yellow HOBBIT T-shirt, decides to join his visitors.

I ask about the oldest members of his crew. Were they digging here already in the 1950s and '60s, 'under Father Verhoeven'?

Theodor Verhoeven requires no introduction – John knows who we're talking about. 'Rokus,' he says. 'But he died last year.' Then he names three more, Donatus, Janto, Benjamin – but none of them is here today. Benjamin, as it turns out, is Benjamin Taurus, the one who found Flo. He left the day before yesterday with the new harvest of fossils, for Ruteng – the town where we had just spent the night.

I'm startled. Don't tell me that yet another 'bone room' has been set up at Hotel Sindha? But no, John tells us, we didn't miss out on something right under our noses: this year they've rented a separate house for storage, in another part of Ruteng.

'Any new *Homo* remains?' Given how sensitive the subject is, I'd been meaning to save that question for last, but my curiosity gets the better of me. The millions being put into the digs at Liang Bua in the form of sponsor contracts can be justified in one fashion and one fashion only: the much-longed-for discoveries of LB10, LB11, LB12 . . .

'Mostly a lot of rat this year,' John says evasively. 'And bat.'

This is no use. I go back to Verhoeven's day: did John know that Liang Bua was once used as a classroom? The foreman grinds out his cigarette butt and kicks it into the pit. 'Not only was this our school,' he says, 'but before our kampong got a church of its own, this was also where mass was held.'

JUST OUTSIDE THE CAVE, back in the present, a sheet covered in coffee beans has been spread out to dry in the sun. The woman in the blue sarong watching over them beckons us to come over. She introduces herself as Theresa, and asks us to follow her to a thatched house surrounded by a wall. In one hand she holds a prayerbook with a threadbare marker, in the other a broom. Two steps ahead of us, Theresa – bent over double as though setting young rice seedlings – sweeps the path clean of twigs and leaves. We find ourselves in the courtyard of the local Liang Bua museum.

While Vera and I are signing the guestbook, we hear our hostess pouring a bucket of water into the toilet. Then we step out onto a tile floor and come face-to-face with Eugène Dubois, from Eijsden. Fancy meeting him here. Dubois is well on in years, balding, dressed in a three-piece suit. His watchchain accentuates the girth of his belly. In Bahasa, the plaque says that he discovered *Pithecanthropus* – later renamed *Homo erectus* – close to Trinil, in Java.

The following panel bears the likeness of the *pastor belanda*, the Dutch priest, Theodor Verhoeven, digging away with his seminary students at Liang Bua in 1965. With Lippus's help, we read that Verhoeven was the first to uncover the remains of elephants and stone tools, which made him surmise that Flores had been inhabited by hominids for the last 750,000 years. 'His insights were rejected, however, because he was considered a layman in the fields of archaeology and geology.'

Turning the final corner, we are suddenly standing before the glass bier of LB1 herself. Cracked skull, lower jaw containing nine teeth, collarbone, right humerus, both ulnae, two fragments of the hip joints, femurs, kneecaps, tibias, metatarsals and phalanges – all on a blanket of black velvet.

'*Homo floresiensis*,' the plastic card reads. 'Female, aged 25–30 years.'

And right above these casts, on the wall:

Welcome, 'Flo'
'Flo' was discovered on 6 September 2003, marking the start of a new
journey through the modern world. Scientists from Indonesia and
other countries are doing a series of studies to reveal her new 'branch'
on the tree of human evolution.

That afternoon, back in Ruteng, we immediately go looking for the
'bone house'. We have been told that it is in the neighbourhood behind
the local landing strip. The problem is, not a single street merchant
or rickshaw driver can point us to it, not even in exchange for a cigarette. But when we decide to just go for lunch at Hotel Sindha first,
our luck changes: the waiter taking our orders tells us that the '*ekska-
vasi* team' had first asked at the hotel about booking a bone room,
but finally decided on a rental house in town. He has no address, but
he's willing to go with us to point it out.

And so, at a quarter past three in the afternoon, with a waiter
in a green hotel uniform as our guide, we knock on the door of the
fossils' guardian. Her name is Rolinda. Blazoned across the expanse
of her T-shirt is the text 'Liang Bua, Home of Civilisation'. 'Oh, but
you just missed them,' she says, clapping her hands in disbelief.

Her housemate, Nona – every bit as well-rounded – comes out
to join us at the garden fence. 'They left at two.'

'They' is Thomas and Jatmiko, the Javanese heads of the excavation team. 'Left' means: they took off on the Trans-Flores road in
their four-wheel drive, on their way to Komodo Airport and a flight
to Jakarta, by way of Bali. Their wooden crates lined with polystyrene
foam contain the season's treasures.

'We just waved goodbye.' To prove it, Rolinda swipes the corresponding photos across the screen of her smartphone. Exactly where

we're standing now, in the shadow of the tamarind, there had been a minibus with huge piles of baggage on the roof and two Indonesians up front. We see them three times: driving away down the street, waving with one hand out the window, turning to look into the camera. In that order. Without meaning to, Rolinda swipes back one photo further into the past. Vera recognises the inside of Liang Bua.

'Yeah, these are from the goodbye party,' Rolinda says. 'We organised a disco.'

On her phone we see the same men, Thomas and Jatmiko, holding plastic plates beside a buffet table. The whole village has turned out in colourful clothing. Speakers are set up close to the pits, which are fenced off with stalks of bamboo. On the dance floor above our ancestral remains, only the foreigners are actually hoofing it – from Hanneke Meijer's description, I recognise a Canadian and an American.

Now that we're here anyway, Rolinda and Nona want to take their pictures with us.

A few minutes later the coffee is boiling on the stove and we're sitting on the couch in the front room, taking selfies, including with the waiter from Hotel Sindha. Lippus is handing out cigarettes.

Coffee here comes with a glass of *arak*, a liquor distilled from palm wine. I propose a toast to the treasurers of Liang Bua, whose work has now been brought to a good end.

Our hostesses glance at each other. They join in the toast, but they're secretive at the same time.

'Our work hasn't come to an end – not yet,' Rolinda says.

Nona chimes in: 'Half the fossils are still here in the house.' She hops to her feet, flapping her little hands: come on, follow me.

Through the kitchen and the living room we enter a hallway leading to a blacked-out bedroom. Lying on the tile floor are the

brittle bones of varans and bats, vultures and storks, dwarf elephants
and giant rats. An orange glow is coming through the lacy curtains
before the single, barred window. From a framed print on the wall,
a benign-looking Jesus watches over all this as he ascends above the
roof of a church.

I recognise vertebrae, parts of jawbones, bird's claws. They are
all labelled: location, finder, date. In a solemn tone, Nona remarks
that no human remains were found this year. The bones are all spread
out on trays. They have been dug from prehistoric times, are not yet
accustomed to the light of AD 2017.

WE ARRIVE AT Leko Lembo much later than we had planned – the
last two hours Lippus drove in the dark. After a long ride downhill,
we smell the sea, hear the rumbling surf.

I'm curious about Janet, and also about her Uncle Thom's experi-
ences with the forest people. Will he call them Iné Wéu or *Ebu Gogo*,
bibet or *koersjati*?

We see Uncle Thom, but cannot talk to him: he has already fallen
asleep. Efi, the helper, has cooked for us – chicken with rice and beans.
After that she went home sick, as she has dengue fever. There's an epidemic at the moment: one out of every five villagers here has dengue.
Janet too. She's recovering now, but the joints in her wrists are still
so sensitive that she can barely shake your hand. Little has changed,
apparently, since 1954, when Father Verhoeven and his students were
forced 'to stop digging [at Liang Toge] due to an epidemic of fever'.

Guesthouse Leko Lembo is a wooden house on piles, with windows and doors – screened off to keep out the mosquitoes, but letting
in the mild sea breeze. A ceiling fan revolves slowly above the dinner

table. During dinner we run through half the world, our shared past, our wanderings. Janet's husband, Sius, is on Java, Lippus is staying with friends, so we can speak Dutch all we like.

Meanwhile, Uncle Thom lies on a little mattress in front of the flickering TV screen, gurgling and murmuring in his sleep. His shirt-tails are hanging out of his trousers.

'Dengue?' I ask.

'*Arak*,' says Janet.

17

BY DAYLIGHT, LEKO LEMBO looks heavenly. Smoke from a woodfire mingles with the tropical dew. At breakfast we eat crepes with papaya slices, while the dogs in the yard play with the washed-up shell of a sea turtle. Hammocks have been slung within ten steps of the verandah. As you rock, you look out over the beach of sand and rocks, an upturned fishing boat, the bend in the bay where the weekly ferry to Sumba docks, and the nose of a volcano sticking out above it all. The look of things is so dazzling that reality makes its way through only with difficulty.

The first rend in this holiday backdrop comes gradually, like a rip in a screen that grows bit by bit. It starts with the sudden shaking of a palm tree – it's as though a mini-whirlwind is rattling only one bit of foliage. The culprit is one of Janet's neighbour's boys, climbing quickly up the trunk. He uses his hands and feet, as well as a girdle of fabric wrapped around both tree and lower back, like a trapeze artist's sling. Swaying his body rhythmically, he more or less walks straight up the trunk, sliding the sling up with him at every step. Applause.

Janet says that he's tapping palm juice for his father. We follow her down a narrow path through the garden plots to the property next door. The woodfire, as it turns out, is part of an improvised distillery. Alcohol is dripping from a bamboo pipe into a jerrycan. The neighbour, who is also the kampong's eldest resident, holds a glass under it. I may (must) taste the moonshine and give my approval.

As we walk on to the only shop in the village, Janet tells us that almost all the farmers here have switched to distilling *arak*. Recently, the stretch of the Trans-Flores beside the bay has come to be called the 'Arak Highway': along both sides are stands selling palm liquor in plastic bottles.

IT SEEMS LIKE those in Leko Lembo kampong who aren't drunk either have dengue fever or malaria. Janet fixes whatever's fixable; she runs all kinds of projects. When the woman behind the counter gives her more change than she should, Janet teaches her how to add and subtract on a notepad. She hands out netting against the malarial mosquitoes, and coordinates the replacement of open-air sewers with properly drained latrines. Via her own foster parenting plan, she has 'adopted' a few of the children, who bunch around to catch a glimpse of us visitors. Their parents are either dead or have moved to Malaysia to find work; a former classmate of ours is the benefactor who pays their tuition.

Apart from the villagers who lie ill in their huts, writhing from the pain in their joints, the fate of the local dogs is the most distressing. Waves of rabies wash over the kampongs from time to time: when that happens, the dogs are beaten to death like minks. The killing is imposed from on high, with no quarter given. During the last canine raid, barely a year ago, Uncle Thom had been forced to dispatch – without the spilling of blood – Janet's favourite bitch, Bobby; in her nightmares, Janet still hears Bobby's shrieks. To make the whole thing even nastier, the neighbours we just met came by afterwards to ask what they were going to do with the dog's body. 'Bury it, of course,' Janet said. But the next morning, she saw that the freshly dug grave had been upturned. Bobby was gone. Whether the animal was infected with rabies or not, she still supplemented the neighbours' dinner.

Janet not only gave us a peek behind the scenes, she also drew us into her family's past – and into that of the Dutch East Indies and Indonesia. The here-and-now might be rough, but that history was even rougher.

As a chaplain in the Royal Netherlands East Indies Army, Janet's grandfather had been taken prisoner on Sumatra in 1942, as had his sixteen-year-old son, Janet's father. They were separated, each of them ending up in a different Japanese camp. Her father's tribulations in the camp had not kept him from travelling later to Sumba as a 'servant of the Word'. In 1956, he and his wife settled as missionaries in the capital town of Waikabubak; they were the only Dutch people there, except for the local doctor and his family.

After her parents' death, Janet had done some research into their lives. Sumba had always been a remote post of interest to no one, where nothing ever happened. That all changed in 1965, the year Janet was born in Waikabubak. The island became awash in a wave of government terror. Her father was picked up twice by the military police. The first time because the grass in the little plot where the missionary family was supposed to fly the Indonesian flag had not been mown properly. The second time – Janet's eldest brother had run crying behind the jeep that took his father away – it was all much grimmer.

When she was cleaning up her parents' home, Janet had found copies of letters describing the gulf of executions that had spilled across the archipelago in 1965 and 1966. To a British historian, Van Oostrum had written that his second arrest came 'in reaction to a rather defiant sermon on Matthew 5:43–44'.

> *Ye have heard that it hath been said, Thou shalt love thy neighbour, and hate thine enemy. But I say unto you, Love your enemies, bless them that curse you, do good to them that hate you, and pray for them which despitefully use you, and persecute you.*

By standing up for the prisoners of Sumba, the preacher had quickly become numbered among their ranks.

Janet read aloud to us from her father's original letters. In them, my old religion teacher cited the commander who came to inform the prisoners of their impending execution: 'Seeing as we are not dealing with animals but with humans, we will not do this [the execution] without pastoral care.' Immediately after this announcement, Van Oostrum was appointed the men's chaplain. The orders to kill these countrymen came from faraway Jakarta, after an aborted military coup on 1 October 1965, in which six generals were killed. President Sukarno's director of military intelligence, a major-general by the name of Suharto, responded by unleashing a purge of the numerous members and sympathisers of the *Partai Komunis Indonesia*. While Suharto was engaged in having as many (supposed) communists as possible exterminated, he also forced President Sukarno to the sidelines and seized power himself.

It was on Java and Bali that the massacres were greatest, but they crossed the Wallace Line too. From the town of Ende on Flores, the army leaders at Waikabubak on Sumba received orders to line up the 'real communists' in front of the firing squad. In the cell complex, Janet's father had celebrated communion with the condemned men. The next morning, he went along with them in the back of the army truck to the beach at Pantai Rua. Right before they were blindfolded, he prayed with them. 'On 5 May and 9 June 1966, I accompanied groups of six and fifteen condemned men to their place of execution,' he wrote.

Until the day he died, Pastor Van Oostrum never spoke to his children about this. It was only on his deathbed that he called his youngest daughter in and told her about the confidential 'letters', and where she could find them: in his study, folded into a volume of the book *Palms and the Cross*.

In 1965 and 1966, at least a quarter of a million Indonesians were put to the sword. Most estimates assume twice that number; the highest ones put it at more than a million.

With carbon copies of the typed letters in my hand, I suddenly felt ashamed for having so often cut Van Oostrum's religion class. It felt as though this was his payback: compared to his testimony, my own quest seemed decadent. What did that one skull from Flores mean in the light of hundreds of thousands of others? Didn't the paltry bones of Liang Bua pale by comparison with the contemporary, still hidden charnel fields of greater Indonesia?

It made me reflect on humans as murderers. *Homo sapiens* was a maker and a breaker, a creature that fouled its own nest, but that also stood out by virtue of its ruthlessness. The exhortation to 'love thy neighbour' had been consistently ignored for the last two thousand years, while the saying that 'man is wolf to man' is an insult to the wolf. Forget the python, forget the megalodon – the most dangerous animal is us.

My daughter and I are out to track down a shark's tooth from the Verhoeven collection and take a sample from it, but what is the use of that? Flores and the other islands in the Indonesian archipel-ago are littered with mass graves. The most recent ones, barely half a century old, lie beneath a slab of heavy taboo. Although grisly stories come bubbling up like cadaverine gas, there is no one who digs for the bones of the modern-day people of Flores. Do murder and its denial go together? Is it typically human to stonewall atrocities?

The biggest bloodbath on Flores took place in 1966 in Maumere, the very town where Verhoeven was teaching at the major seminary. Latin and Greek. What had he noticed of the barbarism?

Janet had never come across Theodor Verhoeven in her father's correspondence. 'The Protestant and Catholic missions were separate worlds.' She told us about a Dutch gentlemen's agreement: Flores for the Catholics, Sumba for the Protestants. During his furloughs, her father always travelled by way of Flores, but it was a transit zone where he was not allowed to proselytise.

When she married, Janet – the reverend's daughter – became Catholic. Just as the iguanas of the Galapagos Islands blend into their surroundings, upon marrying she too had taken on the protective colouration of the local majority.

'Well, what about it?' I ask her. 'Do you notice any difference?'

'Mmm-hmm, with the Catholics it's all about appearances,' she says. 'As long as you attend mass, you can go ahead and worship your ancestors as much as you want in your own home.'

Janet's in-laws had expected her to honour tradition by letting herself be locked up in Leko Lembo's longhouse for seven days. During that week, brides-to-be are instructed in marital commitments.

'But I bought off the obligation,' Janet says. Before we can even ask how, she adds: 'With palm wine.'

18

GIGI HIU, LIPPUS instructs us, is Indonesian for 'shark's tooth'. Vera and I practise for as long as it takes us to pronounce it perfectly. As a bonus, we get *gigi kilat* along with it, which means 'lightning bolt' – literally, 'the tooth of the lightning'. Lippus glances at his mirrors, his baseball cap pulled low above his eyes; shifting down, he passes an overpopulated minibus. *Gigi kilat*, he continues, is also what one calls a meteorite. The shaman in his village has a tiny piece of one. 'He carries it around as an amulet. It's a lucky stone.'

Lippus is kind enough to add a prophecy of his own to this explanation: he says he has a premonition that our mission will succeed. Then he flicks up the sun visor with a quick wave of the hand, like a knight homing in on his opponent on the tournament field.

During our three-day journey to Maumere, we visit three spots where Verhoeven's fossils might be stored or still on view. Our sole, flimsy lead is a photo of a yellow shark tooth against a sea-green background, taken on Flores in the 1970s. '*Collectie Verhoeven*' is printed on the back – that's all we have to go on. A couple of grams of enamel would be enough to date it, a tiny section a few millimetres thick. There's no mention of where the find was made, but José suspects it was dug up at Liang Bua. In combination with its age, it could also shed light on the original marine environment during the period when the mountain ridges of Flores rose up from the sea.

Past the stands of the palm-wine vendors, the Trans-Flores road climbs through eucalyptus forest to a plateau. The island's backbone, Lippus tells us, is a chain of nine volcanoes, the two most active of which are both close to here. We cut around the grey ash cone of Mount Inerie. Somewhere in this sparsely vegetated landscape, between the basalt blocks, must be the cave where the villagers of Ola Bula kampong once exterminated a community of *Ebu Gogo* in a blaze of five hundred palm branches. When it comes to butchery, no moral progress has been made since. Ola Bula no longer exists, but the cave does. There may no longer be maggots crawling from it, but they say you can still see a charred spot on the ceiling.

We drive on to the retreat of the Styler priests at Mataloko, Father Verhoeven's first post, and the place from where he set out in the 1950s to do his digs. On the basis of his celebrated discoveries (he was invited in 1957 to hold a talk at the villa of one of Sukarno's cabinet ministers), Verhoeven had hoped to 'spend more time on prehistoric studies'. In the notes he left behind, he writes about the refusal he met with: 'The provincial abbot, however, saw this as a useless waste of time. His comment, quite literally, was: "What would the other brothers say?"'

Mataloko stretches out in front of us like a ribbon of clutter, a succession of little shops, roadside stands, motorised rickshaws and gas stations. The grounds of the church itself form a strict and sober contrast: they are trimmed and raked in thoroughly un-Floresian fashion.

Behind an ornamental fence lie closely mown lawns, with conifers straight as candles and spherical boxwoods. As we walk towards the main building, we come across no one; this is a sanctuary where every voice has fallen still. Where's the life around here? Father Verhoeven resided in this sanctum for almost ten years. He put his finds out on display in one of the annexes – some of them, or perhaps a list of them, might still be lying around.

Looming over the oval flowerbed before the entrance is an ashen-grey sculpture of 'St Arnoldus Janssen'. The founder of the *Stadt Gottes* has a high forehead and slicked-back hair, his right hand rests on his heart. What a sad, misplaced figure. Why didn't he keep his disciples behind the walls of his monastery on the Meuse? Sorry, Father Verhoeven, sorry, Pastor Van Oostrum, but for the moment I can't view the mission work of Christianity as anything but the continuation of colonialism by other means. Where does this human urge to proselytise come from?

When a gardener steps out of a rose bed, pruning shears in hand, we hold up the photo of the shark's tooth for him to see. But it's only for show. I want to get out of here.

A little later, as we set out on our journey again, I try to explain my distaste. It's not all that complicated: on the verandah this morning I read, in nauseating detail, about the role played by the Catholic clergy during the great massacres of 1965 and 1966. If only turning a blind eye had been the extent of it.

As well as her father's letters, Janet also showed me a chronicle, the only one of its kind, dealing with the killings on Flores throughout three months of the year 1966. 'The Silent Scream of a Silenced History' was the title of the exposé. The author was in fact a Steyler missionary himself, an Englishman by the name of John Prior. As an inquisitive member of the SVD, based in Maumere, Father Prior had written to his elderly predecessors to ask what they remembered of that orgy of violence. His respondents were Dutch priests in retirement, spending their twilight hours in the monastery back at Steyl.

'Memory is just about everything we are,' was John Prior's opening sentence. 'What else is history but a process of remembering?' Those who remembered had a past, those who learned from it had a future – that was Father Prior's motivation. The capacity to remember, in his view, was very close to what it means to be human.

It made a certain sense to me to consider humankind's capacity to hold on to memories as something exceptional. As adept as elephants (and dolphins) are at remembering things – and, who knows, perhaps less selectively than us – it is only we *Homo sapiens* who save our memories in archives and libraries. That knowledge has helped humans to ponder our (mis)deeds and, as the case may be, to commemorate our victims. Along with John Prior, one might even believe that people are capable of learning from their mistakes.

As I absorbed the raw facts from 'The Silent Scream of a Silenced History', I had at times to force myself to keep reading. When it came to gruesomeness, John Prior's true crime report beat Raymond Dart's 'killer ape' hypothesis hands down. Later, I found more first-hand recollections, registered in the words of priests who had been there.

Antoon Bakker from Wervershoof:

Here on Flores, the executions began in late December of 1965 . . .
That meant being clubbed to death, stabbed, kneeling by their grave
and dumped into it by a bullet. For those of us who had to witness
this from close by, it was nerve-racking. The local people were herded
together to watch. We missionaries did not have to do that.

This took place behind the cathedral at Ruteng, which today still stands out in immaculate whiteness against the background of green hills. While looking for Rolinda and Nona's 'bone house', we had driven past it (white with two red, pointed roofs) a few times.

> *In the afternoon, those who had been taken prisoner were transported in trucks to the cemetery. [There] they were shot dead. That took place day after day, and whenever I heard the trucks again I would go to the big cathedral, to that big, big cathedral, and there I would pace back and forth. Bang, bang, bang, bang, bang.*

Along the very road we are on now, Father Bakker had stopped in at a parish where he, in his younger years, had converted almost all the inhabitants. In a shed at the edge of the village, he heard, 'some twenty Communists had been locked up, and tortured considerably already'. He received permission to visit them. They were, as it turned out, not communists but a group of villagers who had joined the local Pentecostalist congregation. Antoon Bakker felt almost no pity for them.

> *As soon as I entered their camp, the former guru Agama began to weep inconsolably, and the others with him. They wanted me to hear their confession, but they were open apostates and I considered them to be in no mortal danger whatsoever. It had been at least twenty years since they were practising Catholics.*

Bakker thanked his lucky stars that he had enough rosaries with him. He handed them out, then went away.

Passing by on my way back, my former cook told me that they had been crushed like serpents, brains squeezed out. When it was the final one's turn, he was still praying the rosary. Just ten more Hail Marys, he said. And they waited for him.

The church archives in the city of Ende showed that the clerical leaders on Flores had encouraged the 'extermination' of all unbelievers and those of other faiths. To his dismay, John Prior had discovered that prominent members of the *Partai Katolik* had taken part fanatically in the round-ups and the killings by torture. From bishops to parishioners, countless Catholics on Flores had proven willing accomplices to the army.

Father Prior cited a letter of instruction from the cardinal of Ende, the highest-ranking Catholic in Flores. During the peak of the reprisals, on 10 March 1966, Monsignor Gabriel Manek addressed his flock with the words:

We thank God that the serpent's poison, which had spread widely in the body of society, is now being rooted out and destroyed. This extermination, by a nation that was threatened by dangerous elements, is nothing more than our obligation to make ourselves secure.

Our religious houses and seminaries should organise a weekly Holy Hour with the usual prayers, Stations of the Cross, and rosary . . . as reparation for the errors and excesses beyond the law of extermination.

Father Verhoeven must have been among those who found this letter in their mailbox. His superior, Father Boumans from Kerkrade, had seen army trucks drive by 'filled with prisoners on their way to be killed'. This took place around midnight on 16 February 1966, right below his bedroom window. At that point, Father Boumans was rector of the major seminary at Ledalero, in the hills above Maumere, where Father Verhoeven taught Latin and Greek. Their rooms lay along the same corridor. For nights on end, the screaming outside kept Father Boumans awake. He heard soldiers egging on the neighbours to curse the victims roundly. Beneath a scourge of maledictions, the 'atheists' had to dig their own grave. As soon as the pit was knee-deep, they were murdered.

'From my room I was only able to kneel down and pray for the victims,' he confessed forty years later in a letter to Father Prior.

The next morning, members of the local Catholic youth movement crowded into the rector's office. They asked Father Boumans to prepare a mass grave for the godless communists on the grounds of the seminary, which were big enough. 'I refused, saying, "That is not possible. This is sacred soil that has been blessed by the Church."'

No one, not even his biographer, seemed to know what Father Verhoeven had seen or heard, and what effect it may have had on him. From Von Koenigswald's archives at the Senckenburg Museum of Natural History in Frankfurt, I had received a stack of scanned letters written by Verhoeven, dated around that time. Each one of them struck a friendly, almost intimate tone. Nowhere did despair shine through, or even a hint of the fact that the population of Flores was in the throes of death.

While here and there the sweetish odour of rotting human flesh hung in the air around Maumere, Verhoeven was busily making plans for new excavations. What bothered him was the refusal of a research grant, which he suspected had something to do with the director of the National Museum of Ethnology in Leiden.

Flores, 10 July 1966

Dear Professor,

My application for a grant from the Dutch Research Council has become stranded, on the advice of Dr Van Heekeren. What reason could he have for doing that?

In a pastoral letter issued in November 1966, after the violence had run its course, the Indonesian bishops closed ranks in support of the *orde baru* ('new order') of General Suharto. Not long afterwards, on Christmas Day, Theodor Verhoeven lost control of the wheel of his jeep and missed a curve.

Two months later, on 25 February 1967, he shows up again in an article in the Dutch daily paper *De Volkskrant*. 'Father Dr. T. Verhoeven (59)', now 'swathed in plaster', is recovering at Rotterdam's St Franciscus Hospital. In the accompanying photo we see him in his hospital pyjamas. Smiling broadly, a twinkle in his eye, he holds up his right arm, in its cast.

The carcass of his mission vehicle was still at the bottom of the ravine. 'They finally got me back up,' Father Verhoeven explained. 'They carried me to the presbytery in a wicker sedan chair.' Like the bones of the long-skulled pygmy of Liang Toge (in their communion tins), this time he himself was lashed to a stretcher and sent off to Surabaya, and from there to the Netherlands. The rest of the newspaper interview is all about fossils. Was *De Volkskrant* simply unable to come up with any more newsworthy questions? Or did Father Verhoeven prefer to steer away from the mass murders – head down, teeth clenched?

Verhoeven was not apolitical. The story went that he knew President Sukarno personally. In any case, they had met once on

Flores, at the opening of an 'in-house museum' in the town of Ende. This 'Bung Karno' house was legendary. In the 1930s, the Dutch authorities, fearful of his oratory skills, sent the militant young Sukarno into exile there. In 1954 he returned to Ende, this time as head of state. The house had been converted into a place of pilgrimage, a commemorative monument on the road to hard-won independence. The only problem was a lack of exhibits: that was why Sukarno had assigned Verhoeven two rooms in which to display his dwarf elephant molars and giant rat skeletons – and probably his shark's tooth as well.

In the Netherlands, the daily newspaper *De Tijd* ran a report of the opening: 'Father Verhoeven showed us the photos of a smiling President Sukarno . . . Among the Indonesians, personal friendship takes priority over politics.'

As it was in 1930, when the mission feature film *Ria Rago* was released, Ende today is a Muslim bridgehead on the south coast of Flores. The town's built-up area stretches across a gently rolling peninsula that has more minarets than church steeples.

An hour before closing time, we get to the museum that Sukarno and Verhoeven outfitted back then with their divergent collections of memorabilia. We're planning to ask whether there is still some corner dedicated to the Verhoeven collection, and if so, where we might find it. But the door is locked and the shutters have been rolled down over the windows. Inside the compound, there is no one in sight. The porter's lodge is empty as well.

On the street corner outside, Lippus approaches a group of *ojek* cabbies. They're lounging around on their motorbikes, waiting for fares. No, we don't want to go anywhere – we want one of them to fetch the concierge, with key and all. After a bit of haggling, two of the motorcycle taxis are willing to try it, for 5000 rupiahs apiece. If they come back with the concierge, or with the museum director, they get double that. The boys race off down the street, each in a different direction.

Meanwhile, one of their colleagues shakes his head. 'It's Saturday afternoon,' he says. 'The start of a three-day weekend. Tomorrow's Sunday, and Monday is International Labor Day.'

We wait for the *ojek* cabbies until closing time. Both of them come back, but they're empty-handed.

19

I N AN EMAIL THAT OPENS WITH the salutation 'Salaam',
John Prior invites us to visit him in Maumere. His seminary
students will be taking their exams next week, but we're welcome in the meantime – 'anytime'. I'm curious to hear his views on
the role of remembrance and commemoration, but also on the human
capacity for forgetting – what does that say about us? Committing
crimes against humanity is one thing, keeping quiet about them is
another. I want to ask Father Prior the same question that he posed
in his exposé: how were we able to 'erase from our collective memory' the horrors of 1965 and 1966?

Maumere is 65 kilometres from Ende, as the crow flies, across
the mountains along the northern coast, but by road it's almost three
times that far. We decide to break our journey in two and spend the
night at a tourist village at the foot of Flores' most popular attraction:
three adjacent crater lakes in three different colours: dark green, red
and turquoise. In 1927, eleven years before Ms Keers passed by with
her measuring stick and skin-colour fan deck, the Dutch governor-
general unveiled a 'monumental bench and flagpole' on the rocky

outcrops between the lakes. There, to the sound of the Dutch national anthem, the tricolour of the Netherlands was raised.

Today, in 2017, the bench and flagpole have been replaced by an Indonesian monument. Along with a crowd of Javanese holiday-makers, we climb a steplike path over the edge of the crater; it ends at a large sign: 'Each year on independence day, the local population gathers here to show its gratitude for the past.'

Lippus confirms what I already suspect: this is where Floresians come each year to commemorate the *proklamasi* of 17 August 1945, with which Sukarno tossed off the yoke of three hundred years of Dutch domination. (He was four years too early – two bloody colonial wars were soon to follow – but still.) The things a natural wonder can bring about.

On the horizon we see the silhouettes of Sumba and East Timor; from right below us, sulphur fumes rise up that can take your breath away when the wind is wrong. One of the Javanese tourists feels like he's choking; another grows dizzy and begins vomiting. The view of the three basins is sublime, the reflection off their surface phenomenal. Anyone who falls in won't drown, but be dissolved by the acid.

'This is the final resting place of the soul,' a second sign says. 'All souls return here when their life's journey is over.'

ON MONDAY MORNING, we report in at the major seminary at Ledalero. The campus is in the mountains above Maumere. We were expecting hectic activity, the exam rush, priests-to-be bustling from one classroom to the other. But the church grounds are as hushed as the Mataloko retreat was. Here too – in the middle of a pond with goldfish – St Arnoldus Janssen watches over things, in exactly the same pose, on a stone pedestal.

We step up onto the concrete verandahs of the dormitories, shout 'hello' and 'salaam'. Then, suddenly, the silence is broken by the sound of a moped pulling up beside us. When the Floresian driver takes off his helmet, we see that he is wearing a clerical collar. 'Today is May Day,' he says. 'Who are you looking for?'

FATHER JOHN PRIOR teaches 'missiology', a subject I have a hard time imagining. The art (or skill?) of proselytising? When we finally arrive at the SVD order's men's quarters back in Maumere, we find the Steyler priest hunkered down in an armchair on the porch, shivering with fever. It's a minor miracle in itself that we have found the place, surrounded as it is by the aerial roots of a tropical tree. The Floresian priest on the moped, it turns out, was recently ordained by Father Prior himself. Driving out in front of us like an *ojek* cabbie, he guided us in ten minutes to this house on the coast, its windows blacked out by heavy curtains.

'This place was under a reign of terror for thirty-one years,' Father Prior says. 'No one dared open their mouth.' A flipflop dangles from his big toe. He runs a hand over his T-shirt, which is printed with images of a few old commemorative stones and the text 'Flores Purba/ Antiek Flores' – the same phrase, 'Ancient Flores', in Indonesian and Dutch. 'And nothing has changed in that regard.'

After General Suharto stepped down in 1998, John Prior had suggested it might be time to erect a monument to the victims of 1965 and 1966. 'That didn't go down well. Someone in power threatened to report me to the army.'

Less than a kilometre from here, between the local parliament building and the harbour, he says, is a shallow, hastily covered pit containing an estimated eight hundred to two thousand bodies. John Prior recently proposed that this mass grave be opened, and the corpses identified by DNA samples and then reinterred in the cemetery. 'That got me a warning too.'

I ask him what that warning involved.

The priest removes his spectacles, rubs his eyes and puts them back on. I can't quite understand his reticence – or is it suspicion? Sitting across from me is the author of a flaming indictment; in comments on the internet, Father Prior has been received as 'the new Luther'. Only when Lippus gets up and goes to his car does Father Prior set aside his distrust. He speaks of a 'wave of blood that washed over the archipelago like a tsunami'.

As soon as he aired the dirty laundry in 2011 with 'The Silent Scream of a Silenced History', his fellow priests started avoiding him. A renewed silence made its entrance. 'Silence again,' the missiologist says. 'No one wants to go digging up the past.'

The most fervent wish of his fellow believers can be summed up in four words: let bygones be bygones. While the curious go on digging at Liang Bua, Maumere wants to keep its boneyards covered at any price. Could it be that we want to know where we come from, but preferably not what we have become?

When I bring up Theo Verhoeven, I do so by way of a neutral object: his shark's tooth.

John Prior looks at the picture. His expression clouds over, although he knows very well where we have to look: in the mission museum, on the campus of the major seminary. There lies all that remains of Father Verhoeven's fossil collection. 'But it's awful.

It's a mess.' The man who watches over the collection, he clarifies, is not interested in archaeology. 'His predecessor was, but he saw the museum as his own little shop and sold the nicest objects under the table. A copper gong? Gone. The big tusk? Gone.'

I feel sorry – in retrospect – for Theo Verhoeven.

Father Prior never met him, he tells me, but he knows that Verhoeven was married to a former nun. He also knows about their honeymoon to Flores, in the late 1970s.

I ask what he knows about Father Verhoeven at the time of the mass murders.

'That he was teaching here. As a scholar of the classics.'

Was it possible that he might not have known about the massacre at Maumere?

'No.'

Was he a witness to it, and did he muffle it away?

'To ask the question is to answer it.'

THE RECENTLY ORDAINED priest who brought us to the SVD house is also willing to track down the curator of the Verhoeven collection. Again, he drives his moped some twenty metres in front of our Toyota, signalling turns left and right with his arms. On our way to the little museum, he's first going to show us Maumere's open secret: the killing fields close to the pier and the mangrove forest.

In the privacy of the car, Lippus drops his guard. 'This has never happened to me before! Normally, I tell my guests what's to see around here. But you people have shown me a side of Flores I never knew.' He knows nothing about the massacres of 1965–66. Never heard about them. That he came into the world during an orgy of violence: Lippus can barely believe it.

That changes, though, when the priest pulls his moped up beside a low stone wall choked with weeds. He signals to us to pull up next to him. 'Don't get out here.' Without removing his helmet, he looks around. Then, with a nod, he indicates the vacant lot on the other side of the wall, where a goat has been put out to graze. Tied to its iron pen with a length of rope, the animal looks at us and chews. The goat is apparently moved from time to time; the weeds have been nibbled short in circles. Beneath those circles, our guide signs to us, lie the bodies.

Two seconds later, he says: 'Let's go!'

I quickly take a few pictures out the car's open window.

'Dad!' Vera says.

Lippus hits the gas. He switches spontaneously to Dutch. '*Godverdomme!*' he swears.

Vera says I shouldn't have taken those pictures. She sounds frantic. 'I don't understand what we're even doing here,' she says.

The priest signals with his hand. We turn left, into the hills. Indeed, I had imagined our trip very differently too. We were going to go in search of little people and huge rats, dwarf elephants and giant storks, of what was considered normal and what was not. I understand that buried beneath the grass at Maumere is the present, and not the looking-glass world of long ago, yet to me Liang Bua and Maumere are two sides of the same coin. But how to explain that? 'Humans are an aberration,' I say. 'Except we've declared ourselves the norm.'

The next moment we find ourselves in heavy traffic. An open-air mass is being held to consecrate a new church that is still under construction. We move at a snail's pace past rows of churchgoers, who are singing and waving palm fronds. The priest on his moped in front of us makes the sign of the cross.

THE MUSEUM WE'RE looking for is housed in a modern building that resembles a chapel. It's closed. Nevertheless, our guide succeeds in finding the concierge. The man is out in the middle of a rice paddy but, miraculously enough, he has mobile phone coverage there. International Labor Day or not, he's on his way.

Half an hour later he swings open the wooden door; a few fluorescent ceiling lights stutter and then glow. The room remains rather dark. In something that resembles a low counter of glass display cases, we see the jawbone of a dwarf elephant. The remains are marked with a card: '*Stegodon sondaari*'. These are the teeth of an indigenous species that was named after Paul Sondaar.

Atop the glass case lies a copy of *Anthropos* magazine, from 1970. It is open to the article in which Father Verhoeven explains that Flores must already have been colonised by hominids 750,000 years ago.

I clap the dust from my hands. Then I ask whether the Verhoeven collection also includes *gigi hiu*.

Right away, the concierge leads us to the back of the room. In a display case there, at eye level in the semi-darkness, is the megalodon tooth from our photograph. But the sliding glass front is equipped with a metal showcase lock, and the key was lost long ago. No worries, though: the concierge pushes the door hard, until it pops out of its rails with the sound of scraping glass. He hands me the shark's tooth; it's so big that I can barely wrap my fingers around it. The tooth is heavy and beautifully striated, with grooves you can barely feel running through the enamel from base to point. I ask if perhaps I could look at it outside, in the light. That's no problem either. We place the tooth on the pavement in front the museum, beside the ruler we brought with us. The fossil, millions of years old, measures 9.8 centimetres in length. Lippus comes for a look, leaning over, hands on his knees.

I ask the concierge whether he might allow us to break off a little piece of it. Between thumb and forefinger, I show him how large – or, rather, how small – that piece would be. He nods politely, walks off and comes back with a knife.

I'm allowed to use the knife myself, to pry off a tiny corner of enamel. Vera films the operation with her mobile phone, and Lippus claps his hands. But I need him to hold open the ziplock bag for me. I drop two splinters of enamel into it. Then we zip it closed carefully. Not long afterwards, as we drive away, Lippus honks the horn just a little too loudly and long. It's the sound of release.

20

O NE SATURDAY IN the late summer of 2017, a year after
my teaching stint, my former student Mariëlle sends me
an email with the subject line: *Tomorrow!*

I've probably already heard, she writes, but tomorrow morning, around church time, the skullcap, thigh bone and molar of the
Javanese ape-man will be on display. Not the replicas that we han-
dled – no, the real thing. Only briefly, from 10 a.m. to 12 p.m., in
Leiden. Exclusively for the participants in an international conference on human evolution.

'Hosted by Naturalis,' says the announcement Mariëlle has
dredged up from the pages of the Human Origins Group website.
I read it with a frown. The Naturalis museum is being renovated and
won't open again for a few years, but the world-famous triptych of
paleoanthropology will still be on display for two hours, as though
it were the Shroud of Turin. 'At the Pesthuis.'

I call José Joordens – she's vacationing on the island of Texel. I
call John de Vos – he's at home. Both of them refer me to the professor of archaeology whose office we used that day to pass around

the replicas. He's the head of the Human Origins Group (and also a great-nephew of Eugène Dubois), and has interrupted his sabbatical specially for the pop-up exhibition. I've met him already; we talked to each other once at an Amsterdam cafe, about the Neanderthals' use of fire. 'I'll see if I can fix it for you,' he says. 'I'll call you back.'

The verdict is reached after a conversation with the security contractor: I'll be given a 'ten-minute slot'. Between 11.50 a.m. and noon, I'll be allowed to make my circuit of the holotype of *Homo erectus*.

FOR THE DUTCH, despite the celebrity of Eugène Dubois and his successor Ralph von Koenigswald, 125 years of skull-hunting had not been an unqualified success. Telling in this regard was that Lucy and the Taung Child enjoyed more fame worldwide than the Javanese ape-man, not only among the general public but also in professional circles. Since 1999, the United Nations' seal of approval adorned the claim that the Cradle of Humankind lay north-west of Johannesburg. No one protested against that, but in that very same year the search for ancient skulls once again moved 'out of Africa'. All eyes suddenly shifted to the Caucasus, to the border area between Europe and Asia. Georgia may not have been the cradle of humankind, but the lowlands between the ranges of the Greater and the Lesser Caucasus were apparently a transfer table, an intersection once used by various types of archaic hominids.

Amid the hayfields and vineyards around the village of Dmanisi lies what is these days one of the world's most productive hominid mines. In a 25-year period, five (non-similar) skulls have been dug up from an old lava bed here, all of them 1.8 million years old and belonging to ... well, to what or to whom, exactly? Let's just say: to *Homo georgicus*.

In that short time, the 'Dmanisi Five' have produced such a sensation that, in the fall of 2016, Georgia hosted an international scientific
congress on the subject. The occasion was represented by the formula
'100 + 25', which signified:

- 25 years since the Georgian paleoanthropologist David
  Lordkipanidze (with his gold Rolex) had discovered the
  first jawbone of *Homo georgicus* (1991)
- 125 years (100 + 25) since Eugène Dubois discovered the
  skullcap of the Javanese ape-man (1891)

And now, precisely one year later, in September 2017, it is Leiden's
turn to receive the community of experts on the protohuman, and
the icing on the cake was the Sunday morning pop-up exhibition.

MARIËLLE AND I agreed to meet up on the west side of the central train station – the 'exact sciences' side. We were betting that she
would be allowed in with me, although that depended on how many
people showed up: officially, she was on the waiting list for the final
ten-minute slot.

  We were early, so we settled down in the sun on a bench in the 'bioscience park', across from the stately Pesthuis. The seventeenth-century
stone plaque above the entrance to what was originally built as a hospital
for victims of the plague showed a fury tearing a child from the arms of
its inconsolable mother. Strange to think that behind these four walls,
at this very moment, the '*Night Watch* of Anthropology' was on display.

  Mariëlle was tanned from a recent sailing jaunt and wore sun-
glasses, which she slid up onto her hair like a diadem with each cloud

that moved before the sun. As soon as I'd thanked her for her tip, I told
her the incredible story I'd heard only the day before from Dubois'
great-nephew: ten years ago, in 2007, Naturalis had been given one
of the five original Dmanisi skulls on temporary loan. 'You'll never
guess what the Georgians got as collateral.'

Mariëlle thought about it.

I told her again that she didn't stand a chance – you'd never dream
up something like this.

'A Rembrandt?' she guessed.

'Two Rembrandts!'

Okay, they were etchings from Rotterdam's Boijmans Van
Beuningen Museum. Not oils from the Rijksmuseum. But still.

'And then, get this: Holland never kept its part of the deal.'

Just as the etchings were about to be shipped to Tbilisi – in
August 2008 – a full-on, five-day war erupted between Georgia and
Russia. The Rembrandts were no longer insurable, and stayed at home
in Rotterdam.

IF ANYONE EMBODIED the hard core of our group of students,
it was Mariëlle. Her ambition was to become a writer, and she still
met regularly with other former students so they could discuss each
other's work. Meanwhile, she had continued to send me a steady
stream of links to the most recent successes in the field of hominid studies. Now she asks whether that multitude of references has
been any help.

Yes and no, I reply. I had been pleasantly surprised by her tip
about how I could make a scale model of the Taung skull roll out of
a 3D printer. 'A skull with a story' was how the vendors advertised

the accompanying software. Their website showed pictures of what the resulting skulls, which included a purple one with dots, could look like.

But there had also been all the hosannas concerning the latest scientific breakthroughs. All those press releases leapfrogging one over the other made me feel like I was constantly lagging behind. If you weren't careful, you could lose your breath.

Barely ten years after the discovery of *Homo floresiensis*, the genus *Homo* had grown another sprig: *Homo naledi*, Star Man, named after the Rising Star caverns close to Swartkrans, in South Africa, where two skeletons were found in 2013. '*Homo naledi* Overturns Conception of Human Evolution' – even the newspaper headlines were being recycled! All the science editors had to do was cut and paste the generic names.

'It's like a soap opera,' I say.

Mariëlle thinks it's more like a Netflix series, which in fact boils down to the same thing. She has hesitated about keeping me up to date on each new episode.

The serial surrounding *Homo naledi* had released a new episode in May 2017, with clickbait for the press: 'Primitive Naledi Man Was Contemporary of Modern Humans'. The Star Man ('almost human') turned out, in fact, to be ten times younger than the original estimation of between two million and three million years old – which meant that he must have lived alongside *Homo sapiens*.

Promptly, in June 2017, the Human Origins Group had come back in the news with another triumph: their visiting professor of paleoanthropology, Jean-Jacques Hublin, was able to attribute a skull found earlier in Morocco to *Homo sapiens* – and as this 'Moroccan' had lived 300,000 years ago, Hublin got away with the scoop that 'we' were now one and a half times older than 'we' had thought.

In the Netherlands, the *De Volkskrant* daily wrote: 'Oldest *Homo sapiens* Ever Lived in Morocco, Not Ethiopia'. 'Just as we had come to the conclusion that our species arose 200,000 years ago in Ethiopia, it now appears that *Homo sapiens* also lived in Morocco a good 100,000 years earlier.'

'Isn't it about time,' I say to Mariëlle, 'that they start announcing the *mistakes* paleoanthropology makes?' With Freek and Roger in mind, I'd started keeping a running list of the age of *Homo sapiens* according to *Homo sapiens*. 'We', as it turned out, had 'aged' in the course of half a century, from 40,000 years in 1960 to 50,000 years in 1980, and on from 200,000 in the year 2000 to 300,000 today. 'If that trend keeps up, we'll be more than a million years old by the end of the century.' I've had a bellyful of the rigorous certainty of our skull interpreters.

'What about you?' Mariëlle asks. 'What's your estimate?'

I say that the only limit I can see lies somewhere in the far reaches of the preposterous.

'So you go along with what the lit professor said?' She's starting to get the impression that I've been converted to postmodernism.

No, I say. The facts won't let me go, and I won't let the facts go. I keep on lifting them, turning them around and examining them under the headlamp of my imagination. But when it comes to what we should understand 'being human' to mean, there are so many theories by now that it's no longer their content that is striking so much as their diversity and ultra-brief lifecycles.

Two incidents – one minor, the other greater – have caused the scales to fall from my eyes. The minor one: archaeologist Dominique Bonjean from Liège, the 'father' of the Child of Sclayn, had seen his claim that 'Neanderthals used make-up' go up in smoke. The head of

Leiden's Human Origins Group, a bigger celebrity than he was, had recently published an alternative explanation: manganese pellets promote combustion; *Homo neanderthalensis* probably collected them from far and near to help build fires.

'Make-up or firelighters...' I say, leaving my sentence incomplete.

A more serious one: after Raymond Dart's death in 1988, his killer-ape story sank into oblivion. In 2006, a new interpretation – with CNN as its propagandist – came up, upsetting the applecart of Dart's old ideas. While out wandering the flats, it said, the poor Taung Child had been seized by an eagle and clawed to death. Scars around the eye sockets corresponded with the injuries that baboons incur when they're killed by raptors. The three-year-old toddler of an ape-man had had his eyes picked out. In this representation of things, the cruel Taung Child, once prototype of the deathly dangerous *Homo sapiens*, had become a victim.

What kind of science is this? Can 'corrections' like this still be ascribed to the advance of scientific insight? Or does the new script simply mesh better with the contemporary zeitgeist? Pop culture has rewritten the scenario over and over again too, every bit as zealously: the makers of the most recent *Planet of the Apes*, for example, caused their hirsute protagonists to evolve in the course of half a century from 'maniacs' (in 1968) to the noblest of creatures (in 2017).

As we cross the lawn to the Pesthuis, I admit to Mariëlle that I almost fell for the Machiavelli thesis advanced by Richard Leakey. I'd enjoyed his *Origins Reconsidered*, from 1992. In it, Leakey goes looking for 'what makes us human', and finds his answer in the power of the human imagination. That which distinguishes us from other species is, as the allusion to Machiavelli suggests, the cunning with which *Homo sapiens* applies the tactics of deception during the hunt. It was not our strength or speed that helped humankind ascend the throne, but our poacher's tricks. Cunning and guile, in other words – those are the traits which formed humans, with which we are cursed and of which we have never succeeded in curing ourselves.

We come to the little drawbridge. On the far side of the duckweed-covered moat we are met by two guards in grey slacks, crease down the legs. The earlier viewings have taken a bit longer than planned, so we are going to have to wait one more time slot. The chances that Mariëlle will be allowed to go in too look better now; in any case, we can wait here, at the gateway to this place of exile, where once citizens infected with the plague were set apart.

Mariëlle wants to know what kept me from embracing Leakey's tale of cunning and guile.

'Dmanisi,' I say.

What I haven't told her yet is that I've just come back from Georgia. I had convinced my wife and daughter to travel with me through the Caucasus during the summer holidays, because I wanted to go to Dmanisi. Each August, an international summer school is held there for a new generation of archaeologists and paleoanthropologists. We arrived on the next-to-last day of the excavation season, and were able to sit in on the wine-drenched closing dinner, with the whole team at one long table beneath the grape arbours.

'In Dmanisi, it's the women who rule the roost,' I tell Mariëlle. The leader of the dig was a female archaeologist whose role model turned out to be Mary Leakey; her assistant was a female archaeologist as well, and her dental bioarchaeologist was a dentist who had given up her practice in Tbilisi because she could no longer stand to listen to her patients whining. 'And three-quarters of the student body were women too.'

Mariëlle asks about the Rolex man – isn't he actually the boss at Dmanisi?

That is true. But Dr Lordkipanidze, a national hero as popular as any Olympic wrestling champion, only showed up when there was something to celebrate, or to lead sponsors around the sites. His watch turned out to have been a gift from Rolex itself, which had also paid for a metal roof over the main pit. 'Before this, we used to work under a tarp,' Dr Teona Shelia had said as she showed us around.

The lava flow on which we had stood stretched out above the rest of the landscape. It was once the lower slope of a volcano, worn away into a narrow ridge by rivers on both sides. Due to its impregnable position, the ridge had been built upon for the last ten centuries: amid the ruins of medieval castles lay gravestones, millstones, pottery. During the Soviet era (in 1983), Lordkipanidze Senior, David's father, had found a few fossilised rhino molars among the amphoras, and then in 1984 several hand axes. After that things got exciting. In 1991, the year of Georgia's independence, Lordkipanidze Junior unearthed from the rugged, pockmarked slope below us the jawbone of a completely unknown hominid. The first two skulls surfaced in 1999, the third in 2000, the fourth (without teeth) in 2003 and the celebrated Skull 5 (currently the most complete) in 2006.

Teona had led us down into what she called 'the champagne room': a shallow cavity where three of the five death's-heads were found. Every time new remains were exhumed, the French ambassador came from Tbilisi to Dmanisi and brought with her a crate of champagne.

Standing on the basalt floor, I'd let the spot work its spell on me. In the spring, on our way to Flores, Vera and I had flown over this place.

Dmanisi lies in a straight, imaginary line between the Netherlands and Indonesia, between the Meuse and the Solo. Here, 1.8 million years ago, there must have been a T-junction: the first (?) primates to leave Africa (?) had seemingly wandered around here amid the sabre-toothed tigers, hyenas and giraffes. Some of their descendants purportedly went west (to Europe), yet others to the east (Asia). Theoretically at least, the latter might have been ancestors to Dubois' Java Man, who, in turn, had perhaps brought forth Flores Man. That this was no bedrock science but only 'a' scenario was something I was doubly aware of at the time. If you heeded the words of Debbie Argue from Canberra, then neither the skulls found at Dmanisi nor those from Java had played a role in Flo's origins. The Australian anthropologist believed that *Homo floresiensis* was a direct relative of *Homo habilis* from Eastern Africa; in interviews, she referred to LB1 as 'the Asiatic sister of *Homo habilis*' – without telling us how they ever happened to end up so far apart.

THAT EVENING IN DMANISI, we'd passed along the platters of grilled peppers, cheese bread, minced lamb and pork shashlik. We drank red wine and salty mineral water from nearby Borjomi. Meanwhile, we talked about hunches and guesses, variations and margins of error.

The Dmanisi Five had generated a whole slew of new mysteries. They were all from the same era, but did not resemble each other. One had a snout, the other's face was flat, number three had a huge underbite, number four had a small head, while number five's head was large. The five 'first Georgians' reminded one of a circus sideshow: five freaks with the only thing in common being how different each was from the other.

We made jokes about that. Enid, from Ethiopia, said that the animals had apparently established a 'people park' here: one of each type, all in their own cages.

'For the giraffes,' chimed in Liv, from Brazil, 'to gape at. That's why they've got such long necks.'

'But seriously,' said Teona. 'If those skulls hadn't all been in the same place, we would have categorised them all as different species, 100 per cent.'

The dig leader got up from the table to toss a few scraps to the stray cats of Dmanisi. She was wearing flip-flops, and her nails were painted black. This was her twenty-seventh year at Dmanisi. In one of the pits here she had met her husband, the father of her children. 'I only missed two seasons.' Teona came back with another carafe of wine, and filled the glasses all round. 'The first time because I was about to give birth, the second because a war broke out.'

She told us about MiG fighters coming in low overhead, in August 2008. Russian killing machines that popped up from behind the mountain ridge, from Armenian territory. Eighty kilometres away, they dropped their payloads close to Gori, Stalin's birthplace. When the first air-raid siren went off, everyone jumped into the main pit, as though it were a trench on the front line. The next day, once the wave of attacks had died down, a line of diplomatic vehicles arrived to evacuate their country's subjects. The summer school participants left for their home countries and continents, and the digging stopped for the rest of the season.

It might have been the hour, or perhaps the alcohol, but at the end of the evening, when I asked what insights the Dmanisi Five have given humanity, Teona threw up her hands. 'All we have are questions.'

I asked her what those questions might be.

'Why we are who we are,' she cried.

Teona's assistant, the geologist Sophia, stopped tuning her guitar for a moment and added: 'Why do we, as a species, exterminate other species?'

The Dmanisi Five maintained radio silence. To us, they had become static, immobile objects, dead matter. They had already traded in their newly given name, *Homo georgicus*, for that of *Homo erectus*. No one could say exactly to which group they belonged, so they were just tossed onto one and the same heap.

Was there, against the background of Dmanisi, nothing sensible we could say, then, about who we are?

Yes, in fact there was. Teona said that one of the five skulls – the toothless one, Skull 4 – had given rise to a new theory. At first it was thought that the skull had belonged to a sick or elderly man, but further study showed that the owner of Skull 4 had lived on for at least two years after losing his teeth. He had been kept alive.

'By his own kind,' Teona clarified. 'They must have brought him soft food. Maybe they even fed him!'

In the silence that fell, I did my best to imagine the Dmanisi folk as caregivers. The thought that humans owed their humanity to caring for the weak was the opposite of the Machiavellian story of cunning and deceit. Teona called it 'a counter-narrative'.

'Do women see things that men don't?' I asked.

'When it comes to scientists, that shouldn't be the case,' she said. 'But in actual practice, yes.'

Teona Shelia said that the 'care for the weak' hypothesis had meanwhile caught on with her colleague María Martinón-Torres, a Spanish archaeologist trained at Dmanisi. She had recently been in the limelight with her publications about the caves of Atapuerca, in Spain. In her study of the misshapen 400,000-year-old skull of a hand-icapped child, María Martinón-Torres came to the conclusion that this girl had not been abandoned after birth. Her teeth showed that she had been somewhere between five and twelve years of age when she died. This disabled Neanderthal child could not have lived that long without the care of adults. Like Dr Shelia of Georgia, Dr Mart-inón-Torres of Spain was now stating that we should see caring for the weak and the ill as the prime, essential characteristic of humanity.

The charm of the idea impressed me. But was there anything to it? Didn't the same, ever-shifting fallacy lie beneath all these explana-tions? It seemed to me no coincidence that insights into reciprocity, the hushing of a baby's cries with speech and the care for the weak had come bubbling to the surface at this particular point in time, now that the first generation of female experts were making their appearance. They seemed to me like highly necessary corrections to established viewpoints, coloured as they had been before by masculine virility, welcome adjustments to a science dominated until recently by Hemingway-esque men. But this did nothing to change the fact that the echoing well was still doing its work. The pitfall of the cliché, apparently, was one no person could avoid.

AT TWO MINUTES to twelve, Mariëlle and I are summoned from the waiting room at the Pesthuis. The guard who comes to fetch us has good news. Of the five candidates on the waiting list for 11.50, three have not shown up. So Mariëlle, a Romanian doctoral student and I would be allowed to follow him through the courtyard garden to the exhibition room.

'Welcome to our treasure-trove.' The new curator of the Dubois Collection – John de Vos's successor – waves her arm to bid us welcome to the tiled room. We enter a space a bit like an auditorium, big and empty except for three display cases. The guard takes up a discreet position against the back wall, next to a colleague and a fire-engine-red extinguisher.

The first display case is cubical and contains, to our surprise, Dubois' etched mussel shell. The zigzag line that José had discovered on it is singled out by an LED light. Even then, you have to focus carefully in order to pick out the pattern. A museum attendant, brought in for the occasion, tells us about it. In a whisper, he says that with her article in *Nature*, José Joordens had demonstrated not only that the grooves were made by hand, but also that she knew what had been used to make them. 'Flint would be a likely candidate, but you don't find that along the River Solo.'

So?

'*Homo erectus* used shark's teeth. You find them everywhere around there.'

I keep my mouth shut about Father Verhoeven's megalodon tooth; in the meantime, we are told that José has dated the shell to seven million years.

We nod and walk on to the second rectangular display case. In its dimensions, it's like a glass coffin. Here lies the one who had drawn the zigzag line.

'Is it a guy or a girl?' the Romanian doctoral student asks.

'If only we knew,' says the archaeology professor – back now from sabbatical.

What are we looking at? Beneath the three-centimetre-thick plate of glass lies the brown skullcap that we have passed around, but this is the real one. It's impossible to tell the difference. Beside it, on a little dais of unreflecting glass: the ape-man molar – complete with bifurcated root and an enamelled crown. Beside that: the left femur, too long and too straight for any quadruped.

My awe wins out over my scepticism. What draws me in is the atmosphere. The silent circumambulation. The aquarium-like coffin. The hemming of the uniformed sentinels. Amid the tolling of bells from Leiden's churches, Mariëlle and I are attending a high mass of science.

Then we come to the third display case. In it lies a sheet of paper with writing on it, under glass as well. It is a yellowed page from Dubois' original manuscript – the report in which he described his discovery. After much thought, he had named him *anthropo-pithecus* (man-ape). But – as the page open before me demonstrates – he had changed his mind. With a firm stroke of his dip pen, he struck out the word. He recognised his error and introduced a crucial improvement: above the diagonal stroke, he scribbled *pithec-anthropus* (ape-man).

The sight of this correction causes me to feel an unexpected shiver. This is *the* revision. As I wonder why this moves me so directly, I realise that, for the last few years, I have been looking not for fossils, but for words. The mussel shell is signed, Dubois' paper written. And I myself am reporting on it – and this is duly noted.

If the search for the human essence has made anything clear to me, it is that we are doomed to go on revising what we think we know. Only fiction can make reality seem to add up – temporarily. All other brands of description will always be open-ended, allowing them to be the start of something new. Version upon version. It is the strike-out, of all things, that sets us apart as a species. The definitive version does not exist.

*   *   *

WHEN OUR TEN minutes are over, the exhibition is dismantled. We linger a bit. From the corner of my eye, I see the two guards attaching hand-sized suction cups to the plate of glass. *Pang*, the plate breaks out of its groove. There is the sound of laughter. One of the Naturalis staff members is wearing a sweater with the announcement 'Fossils Are Fun' knitted into it.

John de Vos's successor, I see, has brought along the original teak-wood box with brass hinges and locks in which Dubois himself kept the skull of his Javanese ape-man. As cautiously as a conjuror, she presses her fingers to the sides of the skullcap, lifts it from the case and arranges it neatly amid the wads of cotton. Then she lowers the lid of the reliquary and closes it without a sound.

Epilogue

AROUND NOON ON 23 MAY 2000, a French scuba diver lowers himself slowly to the floor of the Mediterranean. He knows, in the fishy world in which he submerges himself, his own precise position. The diver is several nautical miles off the coast south of Marseille, not far from a few rocky, uninhabited islands, L'archipel de Riou.

Here the Mediterranean Sea is 230 feet deep. The distance between the water's surface and the sandy seafloor therefore equals the length of 230 shoes, European size 47. About seventy metres.

The diver knows not only where he is, but also what he is diving for: the wreckage of an airplane from World War II. A Lockheed F-5 Lightning, a twin-propellor reconnaissance plane, without guns. It took off on 31 July 1944 from Bogo base on Corsica, and never came back.

The diver is searching with information from a fisherman who, two years earlier, at this very spot, caught a silver bracelet in his nets. On it, in capital letters, a name was engraved: 'Antoine de Saint-Exupéry'. The diver combs the seabed, working in long swathes. After an hour or so he comes across a piece of metal sticking up from the sand, overgrown with seaweed. Through his goggles, he recognises in this chunk of rust the landing gears of the Lockheed F-5 Lightning.

ANTOINE DE SAINT-EXUPÉRY was forty-four years old when he took to the skies and vanished. The discovery of his seaman's grave takes place one month before he would have turned 100.

Under commission from the French Ministry of Culture, a team of divers recovers the wreck. With their iron lungs and feet of webbed rubber, they go swarming out like marine animals. The pieces of wreckage they collect lie spread over a plot 400 metres wide and 1000 metres long.

The find of an aluminium strip with serial number 2734L punched into it provides 100 per cent proof that the plane belonged to the world's most famous pilot author. A reconstruction shows that it must have fallen from the sky 'almost vertically' on that fateful 31 July, and hit the water's surface at a speed of approximately 800 kilometres per hour.

No bullet holes were found. No remains either. It is too late to know whether Antoine de Saint-Exupéry was slung from his cockpit on impact, or if his body was later eased from its straps and washed away by the current.

Sources &
Acknowledgments

I N WRITING THIS BOOK, I was able to draw upon a wealth
of sources. Some are mentioned explicitly in the text, where
necessary. But the lion's share of the literature consulted I have
left unmentioned, for reasons of readability.

From *De grote wereld*, Arthur Japin's contribution to the 2006
Dutch Book Week, I derived a few observations about being small
and about the child's perspective; I included them here because they
illustrate precisely what I wanted to discuss.

The sentence 'Man is an animal that has broken free of the animal
kingdom' is one of the central tenets from the work of the philosophical anthropologist Helmuth Plessner (1892–1985).

The French scuba diver who in 2000 found at the bottom of the
Mediterranean the wrecked plane of the creator of *The Little Prince*
is Luc Vanrell; he is the owner of a diving school in Marseille.

Important background information concerning the life and work
of Eugène Dubois is found in the doctoral thesis by Bert Theunissen,
Eugène Dubois en de aapmens van Java (Rodopi, Amsterdam, 1985).
A few salient details concerning his personality and idiosyncrasies

I found in Pat Shipman's biography, *The Man Who Found the Missing Link: The Extraordinary Life of Eugène Dubois* (Weidenfeld, 2001). Paul C.H. Albers and John de Vos collected unique (photographic) material in the publication *Through Eugène Dubois' Eyes: Stills of a Turbulent Life* (Brill, Leiden, 2010). Dubois' youth in Eijsden is described in detail in the special October 2010 'Eugène Dubois' edition of the magazine *Uit Eijsdens verleden*, published by the Eijsden Historical Foundation. During the 1983 centenary celebration of Dubois' discovery of his 'pithecanthropus', two publications appeared, both with the name of the accompanying exhibition: *Man-Ape, Ape-Man*. The first was written by Mary Bouquet (National Museum of Natural History, Leiden, 1993), the second by Richard E. Leakey and L. Jan Slikkerveer (The Netherlands Foundation for Kenya Wildlife Service, Leiden, 1993). Details concerning the delusions suffered by Dubois at a later age came from *De Leidse Universiteit (1928–1946)* by P.J. Idenburg (Universitaire Pers Leiden, The Hague, 1978).

In addition to Dubois' original publication, *Pithecanthropus erectus. Eine menschenähnliche Übergangsform aus Java* (Batavia, Landsdrukkerij, 1894), I also consulted that of Ralph von Koenigswald: *Neue Pithecanthropus-Funde 1936–1938* (Batavia, Landsdrukkerij, 1940), as well as P.C. Schmerling's main opus: *Recherches sur les ossements fossiles découverts dans les cavernes de la Province de Liège* (L'Université de Liège, Liège, 1833). Everything about the 'Child of Sclayn' can be found in the book, co-edited by Dominique Bonjean, *The Scladina I-4A Juvenile Neanderthal: Palaeoanthropology and Context* (ERAUL Editions, Liège, 2014).

The Von Koenigswald memoirs mentioned in the text are from *Speurtocht in de prehistorie. Ontmoetingen met onze voorouders* (Het

Spectrum, Utrecht, 1962). I found biographical information about Paul Sondaar in publications including *Elephants Have a Snorkel! Papers in Honour of Paul Y. Sondaar*, compiled by Jelle W.F. Reumer and John de Vos (Deinsea, Jaarbericht van het Natuurmuseum Rotterdam, number 7, 1999).

Raymond Dart's student Phillip V. Tobias wrote extensively about his mentor in *Images of Humanity* (Ashanti Publishing, Rivonia, 1991) and *Into the Past* (Picador Africa, Johannesburg, 2005). Lydia Pyne's *Seven Skeletons* (Viking, New York, 2016) deals with Dart's Taung Child and other famous skulls (including LB1).

Most of Father Theodor Verhoeven's writings and notes are kept in the university library at Leiden, in the Special Collections section, catalogue number H 1429. To place him in a broader context, I was greatly served by three publications about the Catholic missionaries on Flores, all by Marie-Antoinette Willemsen: *Bewogen missie. Het gebruik van het medium film door Nederlandse kloostergemeenschappen* (Verloren, Hilversum, 2012, with Joost van Vugt), *Een pionier op Flores. Jilis Verheijen (1908–1997), missionaris en onderzoeker* (Walburgpers, Zutphen, 2006) and *De lange weg naar Nusa Tenggara. Spanningsvelden in een missiegebied* (Verloren, Hilversum, 2015). From this latter work I cited a few reports from those missionaries concerning the killings on Flores in 1965–66. They form a unique addition to the documentation assembled by John Prior in his article 'The Silent Scream of a Silenced History' in the magazine *Exchange* (Brill, Leiden, number 40, 2011). This same period was also described by Lambert J. Giebels in *De stille genocide* (Bert Bakker, Amsterdam, 2005). Paul Webb has written specifically about the role of the Catholic Church in these atrocities in *Palms and the Cross* (James Cook University Press, North Queensland, 1986), in which he also mentions Reverend van Oostrum's arrest. The letters shown me by Jeanette van Oostrum were written in 1987 as specifications (addressed to Webb), which her father did not dare to make public at the time. The Indonesian theologist Mery Kolimon, in *Forbidden Memories* (Monash University Publishing, Clayton, Victoria, 2015), also mentions the courageous role played by Reverend Van Oostrum during his forced presence at the executions of (putative) communists.

AMONG THE SOURCES consulted on archaeology and (paleo) anthropology on Java and Flores were, in addition to *The Negritos of the Eastern Little Sunda Islands* by Wilhelmina Keers (Het Indisch Instituut, Afdeling Volkenkunde, Mededeling 26, Amsterdam, 1948), also the novel *Ria Rago* by the missionary Father P. Heerkens (Het Poirtersfonds, Eindhoven, 1938), *De onmeetbare mens. Schedels, ras en wetenschap in Nederlands-Indië* by Fenneke Sysling (Vantilt, Nijmegen, 2015) and the travelogue *Varanen, orang-oetans en paradijsvogels. Reizen met Alfred Russel Wallace door Indonesië* by Alexander Reeuwijk (Kleine Uil, Groningen, 2018).

Dedicated in its entirety to Flores Man is the book *A New Human: The Startling Discovery and Strange Story of the 'Hobbits' of Flores, Indonesia*, written by Mike Morwood along with Penny Oosterzee (Smithsonian Books, New York, 2007). Apart from the scientific literature, significant in this context are *The Fossil Chronicles* by Dean Falk (University of California Press, Berkeley, 2011), as well as the 'counter-book' *The Hobbit Trap: How New Species Are Invented*, by Maciej Henneberg, Robert B. Eckhardt and John Schofield (Left Coast Press, Walnut Creek, California, 2011). In his article 'Receiving an Ancestor in the Phylogenetic Tree' (*Journal of the History of*

Biology, number 42, 2009), John de Vos intervened in this discussion as well.

Among the other works that influenced my thinking concerning human evolution were *Origins Reconsidered: In Search of What Makes Us Human* by Richard E. Leakey and Roger Lewin (Little, Brown, London, 1993), *A View to a Death in the Morning: Hunting and Nature Through History* by Matt Cartmill (Harvard University Press, Cambridge, Massachusetts, 1993), *The Fossil Trail: How We Know What We Think We Know About Human Evolution* by Ian Tattersall (Oxford University Press, Oxford, 1995) and *Catching Fire: How Cooking Made Us Human* by Richard Wrangham (Profile Books, London, 2009).

Important connections with this theme from the viewpoint of ethology have been established in unparalleled fashion by the primatologist Frans de Waal. His pioneering *Chimpanzee Politics: Power and Sex among Apes* (in which the male chimps of Burgers' Zoo in the town of Arnhem display the most ruthless side of their nature) appeared in 1982, followed by works including *Are We Smart Enough to Know How Smart Animals Are?* (W.W. Norton, New York, 2016). In his *Sapiens: A Brief History of Humankind* (Harper, New York, 2014), Yuval Noah Harari makes a brave attempt to bring together a multitude of scientific facts and insights within a common framework. I read *Sapiens* in the summer of 2017 during my trip to Dmanisi, Georgia. There, beside the digs, I also read *De geest uit de fles. Hoe de moderne mens werd wie hij is* (Lemniscaat, Rotterdam, 2017), in which Dutch philosopher Ger Groot convincingly sheds light on the loss of certainties and not-knowing as typical, contemporary human traits.

I was also able to whet my own thoughts with the reading of *Het geniale dier. Een andere antropologie* by René ten Bos (Boom,

Amsterdam, 2008). In a different fashion, the same applied to Piet de Rooy's *Op zoek naar volmaaktheid. H.M. Bernelot Moens en het mysterie van afkomst en toekomst* (De Haan, Houten, 1991), as well as to the novel *Het tegenovergestelde van een mens* by Lieke Marsman (Atlas Contact, Amsterdam, 2017), the essay *De soldaat was een dolfijn* by Eva Meijer (Cossee, Amsterdam, 2017), *De geschiedenis van de vooruitgang* by Rutger Bregman (De Bezige Bij, Amsterdam, 2013) and *Over oude wegen. Een reis door de geschiedenis van Europa* by Mathijs Deen (Thomas Rap, Amsterdam, 2018).

In addition to *Narratives of Human Evolution* by Misia Landau (Yale University Press, New Haven, 1991), I am indebted to *Explaining Human Origins: Myth, Imagination and Conjecture*, a study in the philosophy of science by Wiktor Stoczkowski (Cambridge University Press, Cambridge, 2002). The work of both these authors is examined critically by Peter J. Bowler in *Studying Human Origins*, a collection with a wide range of noteworthy contributions edited by Raymond Corbey and Wil Roebroeks (Amsterdam University Press, Amsterdam, 2001). Particularly provocative in this context too was *Primate Visions: Gender, Race and Nature in the World of Modern Science* by Donna Haraway (Routledge, Abingdon, 1989). I was fascinated beyond bounds by Landau's application of the findings of Vladimir Propp, from his classic literature analysis *Morphology of the Folk Tale* (University of Texas Press, Austin, 2013).

AFTER COMPLETING HIS commission as the first writer-in-residence at Leiden University, Gerard Reve collected his public lectures in *Zelf schrijver worden* (Sdu, The Hague, 1985). In one of my public lectures, I posited my own ideas on the subject in 'Too

True to be Good: A Response to Reve' (Albert Verweylezing, 2016, published in NRC *Handelsblad*, 4 November 2016), from which a few elements in this book were taken.

I would like to extend my heartfelt gratitude to Professor Yra van Dijk for her supervision of my residence. On any number of occasions, our discussions continued afterwards in her office in the Van Eyck faculty building. I have fond memories of those conversations – Yra was a marvellous coach, both in morale-boosting and with regard to substantive issues related to teaching. I would also like to thank her colleagues Esther Op de Beek and Jaap de Jong, and in particular Korrie Korevaart, without whose decisive action I would have wandered aimlessly about Leiden.

On the other side of the tracks, I received a heart-warming welcome from José Joordens – who these days holds the Naturalis Dubois Chair in Hominin Paleoecology and Evolution at the University of Maastricht. She not only saw to it that my students and I could become acquainted with the casts of the skulls in question, but also enabled a number of important contacts for me. First of all, with 'elder statesman' John de Vos, to whom I am exceeding grateful for his lovely anecdotes and his acuity of mind. It was a great pleasure to share his company, both at Leiden and in Eijsden.

A special word of thanks goes out to the board of the Dubois Fellowship as a whole, and to Ludo Hellemans and Jean Pierre de Warrimont in particular. I would also like to thank Gerard 'Sjra' van Horne of Haelen for the shared glimpses of the history of the De Bedelaar estate.

José Joordens also directed me to Dominique Bonjean of the University of Liège. I especially appreciated our conversation and the tour he gave me of the Scladina cavern. The same goes for the spontaneous and warm welcome I received in the home of Wilhelmina (Mientje) Gérard-van Loon in Awirs.

During a get-together in an Amsterdam cafe, Wil Roebroeks, head of the Human Origins Group of the Department of Archaeology at Leiden University, was kind enough to introduce me to the major discussions going on within his field. I sincerely thank Hanneke Meijer for the candour with which she told me about the ongoing excavations on Flores.

With Harold Berghuis, fellow son of the city of Assen, I hit it off immediately from the moment we first shook hands. In addition to a treasure-trove of information, Harold also provided me with a CD of his jazz band, Hot Club de Frank. My gratitude is great. It also pleases me to say that he finally found a home for his PhD project in the course of 2018, and in Leiden at that, in cooperation with Naturalis.

I also owe a debt of gratitude to Gert Knepper for his revelations concerning his former Greek teacher.

First by means of email and text message, then in real life, Jeanette van Oostrum provided me with invaluable advice on Flores. My daughter and I spent an unforgettable time in her guesthouse on the beach at Leko Lembo, where her sister Ineke was staying as well. On our trips from Labuan Bajo to Maumere and back again, we enjoyed Jeanette and Ineke's hospitality for several days. Much thanks for all the conversation, the tour of the kampong and the contact with 'Lippus', Philippus Neto, who proved to be so much more than a driver: he was a guide and a travelling companion. I am extremely grateful to Lippus for his stories, humour and friendship.

I do not envy Father John Prior of Maumere, but I appreciate him all the more: he is a whistle-blower who is exposed to great personal risks. I am grateful to him for our reception at Maumere, and for the candid email correspondence we carried out afterwards.

At sea I was welcomed in the grandest fashion by the 'Team Van Oord', led by project supervisor Erwin van den Bergh. It was a fantastic experience for me to be able to sign on aboard the *HAM316*. Thank you, Erwin, for our conversations on land in Lincolnshire. Thanks also to Peter Stout, Rowan Piek, Virgil Konings and especially to 'ship's engineer' Wim Balvert, my personal pilot at this dredging concern.

THIS TOO IS the place to bestow the victor's laurels on the students who were my companions during the first three months of this book's life. Those early days were crucial. At such an early stage, a story may take off in any direction, but the writer goes in search of a vector: both direction *and* magnitude. As the winter approached, I gradually came to see all of you, my students, as the members of a bobsled team. The sled had to be equipped and made ready, the runners mounted underneath, the bolts fastened, the cowling polished. When the moment arrived for me to proceed on my own, all of you provided me with a gigantic push – that, at least, is how I experienced our work together.

A very sincere word of thanks therefore to Julia Duijvekam, Ruben van Gaalen, Els Goddijn, Pien 't Hart, Lian Hof, Juul Klein Wolterink, Marijn Klok, Astrid Koopman, Thom van Leuveren, Roger Louwers, Annebeth Maarsman, Bob Pierik, Monica Preller, Jannie van der Reek, Manola Ruff, Hidde Slotboom, Caroline Snoek, Jacqueline Verheijen, Freek van Vliet and Elfrieda Westerbeek. Together they constituted the 'hard core'. Another student who definitely deserves that qualification is Mariëlle Selser. With the following difference: even after the series of lectures was finished, she remained actively involved in the background as this book arose, up to and including its editing. I profited a great deal from her keen running commentary on various versions of the manuscript.

I thank my proofreaders for the breadcrumb trail of indispensable suggestions they left behind for me in the margins: Josje Kraamer, my editor, who succeeds wonderfully well in achieving the finest weave of experience and dedication, and Annette Portegies, my publisher, who put her trust in this book from the very start, and who, regardless of the situation, continued to cheer me on through all the curves with her warmth and professionalism.

To the whole team at Querido Fosfor – Hugo van Doornum, Patricia de Groot, Paulien Loerts, Jesse Hoek, Jolanda van Dijk, Esther van Dijk, Vincent Schmitz, Inge Janssens – and definitely to my agent, Dorine Holman, as well. Thank you!

As with my earlier books, the sharp eye with which Suzanna Jansen scrutinised and commented on my manuscript on more than one occasion was unparalleled. To say that I am delighted with her commitment would be an understatement.

Finally, I'd like to thank Vera Westerman for the originality of her insights, with which she constantly draws my attention to things that would otherwise – to paraphrase Gerard Reve – remain unnoticed.

Amsterdam,
12 September 2018

ABOUT THE AUTHORS

FRANK WESTERMAN is a highly acclaimed Dutch non-fiction writer. His work has been translated into sixteen languages and has received numerous accolades, including the Kapuściński Prize (Poland), the Premio Terzani (Italy) and the Prix du Livre du Réel (France).

SAM GARRETT is an award-winning translator of over fifty novels and works of non-fiction. He is the only translator to have twice won the British Society of Authors' Vondel Prize for Dutch–English translation.